A Beginner's Course in Boundary Element Methods

Whye-Teong Ang

Universal Publishers
Boca Raton, Florida

A Beginner's Course in Boundary Element Methods

Universal Publishers
Boca Raton, Florida • USA
2007

ISBN: 1-58112-974-2
13-ISBN: 978-1-58112-974-8

www.universal-publishers.com

Preface

During the last few decades, the boundary element method, also known as the boundary integral equation method or boundary integral method, has gradually evolved to become one of the few widely used numerical techniques for solving boundary value problems in engineering and physical sciences. In implementing the method, only the boundary of the solution domain has to be discretized into elements. In the case of a two-dimensional problem, this is really easy to do. Put closely packed points on the boundary (a curve) and join up two consecutive neighboring points to form straight line elements.

In March 1985, when I started research work for a doctoral degree in the Department of Applied Mathematics at the University of Adelaide, Australia, I was introduced to the method by my thesis supervisor, David L Clements. At that time, the term "boundary element method" was relatively new. It was apparently first used in a 1977 paper by CA Brebbia and J Dominguez[1]. Carlos Brebbia and his co-researchers had undoubtedly played an important role in introducing the method to the engineering research community. Apparently less than 200 journal papers whose titles contained the term "boundary element method" could be found in 1985. In 2006, there were several thousand or perhaps even more such papers.

The history of the method could, however, be traced back to an earlier time, well before the 1970s. The mathematics that laid the theoretical foundation for the development of the method could be found in the works of famous mathematicians like Laplace, Gauss, Green, Fredholm, Betti, Somigliana, Muskhelishvili, Mikhlin and Kupradze. In the 1960s, there were attempts at using electronic computers to approximate solutions of potential problems through the use of boundary integral equations, notably the pioneering works of MA Jaswon and

---

[1]CA Brebbia and J Dominguez, "Boundary element methods for potential problems," *Applied Mathematical Modelling*, Volume 1, 1977, pp. 372-378.

GT Symm[2]. The work of Frank J Rizzo[3] was regarded by many researchers as the beginning of a novel direct boundary integral method for the numerical solution of elasticity problems.

After completing my doctoral work in the middle of 1987, I continued to keep myself informed on the development of the boundary integral method and related mathematical works, pick up some new ideas now and then, attend conferences, give talks and seminars, and contribute to boundary element research with applications to problems in engineering and physical sciences. Some specific research areas I had worked on using the boundary integral method include linear fracture mechanics (accurate computation of stress intensity factors using special Green's functions), analyses of nonhomogeneous media (such as functionally graded materials), diffusion with specification of mass, modeling of photonic crystal fibers, integral formulation of imperfect interfaces and bioheat transfer.

Sometimes, I undertook the task of introducing the method to beginners, mainly advanced undergraduate and research students who were working on projects under my supervision. To do this, I had produced various notes over a period of time. The chapters in this book were written based on those notes. In writing this book, I assume that the readers have some prior basic knowledge of vector calculus (covering topics such as line, surface and volume integrals and the various integral theorems), ordinary and partial differential equations, complex variables and computer programming.

FORTRAN 77 codes for the numerical procedures discussed are listed in the chapters[4]. Some justifications, if any is needed

---

[2]One may refer to the following papers: (a) MA Jaswon, "An integral equation method in potential theory I," *Proceedings of the Royal Society of London Series A*, Volume 275, 1963, pp. 23-32, and (b) GT Symm, "An integral equation method in potential theory II," *Proceedings of the Royal Society of London Series A*, Volume 275, 1963, pp. 33-46.

[3]FJ Rizzo, "An integral equation approach to boundary value problems of classical elastostatics," *Quarterly of Applied Mathematics*, Volume 25, 1967, pp. 83-95. This was the work presented by FJ Rizzo in his doctoral dissertation. Much later on in 1993, it won him an ASME Warner Medal.

[4]Readers who are interested in obtaining the codes in electronic form may e-mail me at mwtang@ntu.edu.sg.

at all, for using good old FORTRAN 77 would be as follows. Firstly, in spite of its seniority, it still remains a powerful "number crunching tool". Secondly, its codes are relatively easy to decipher and would be of some use even to readers who are attempting to implement the numerical procedures using newer software tools (such as C++ and MATLAB). Thirdly, free FORTRAN 77 compilers (e.g. FTN77 from Salford Software and GNU Fortran) can be downloaded from the internet.

The constant encouragement and support of my dear wife, Young-Soon, had greatly motivated me to start and finish writing this book. I would like to thank Ean-Hin Ooi, Lukito Jayaputra, Bao Ing Yun, Jackson R Jones, Joris Vankerschaver and Alessandro Vaccari for informing me of errors in an earlier version of this book and Jeff Young and Rebekah Galy of Universal Publishers for their prompt replies to all my questions on the publication of this book.

Whye-Teong Ang, Singapore, 2007, 2014

Contents

Chapter 1

Two-dimensional Laplace's Equation

1.1 Introduction

Perhaps a good starting point for introducing boundary element methods is through solving boundary value problems governed by the two-dimensional Laplace's equation

$$\frac{\partial^2 \phi}{\partial x^2} + \frac{\partial^2 \phi}{\partial y^2} = 0. \tag{1.1}$$

The Laplace's equation occurs in the formulation of problems in many diverse fields of studies in engineering and physical sciences, such as thermostatics, elastostatics, electrostatics, magnetostatics, ideal fluid flow and flow in porous media.

An interior boundary value problem which is of practical interest requires solving Eq. (1.1) in the two-dimensional region R (on the Oxy plane) bounded by a simple closed curve C subject to the boundary conditions

$$\phi = f_1(x,y) \text{ for } (x,y) \in C_1,$$
$$\frac{\partial \phi}{\partial n} = f_2(x,y) \text{ for } (x,y) \in C_2, \tag{1.2}$$

where f_1 and f_2 are suitably prescribed functions and C_1 and C_2 are non-intersecting curves such that $C_1 \cup C_2 = C$. Refer to Figure 1.1 for a geometrical sketch of the problem.

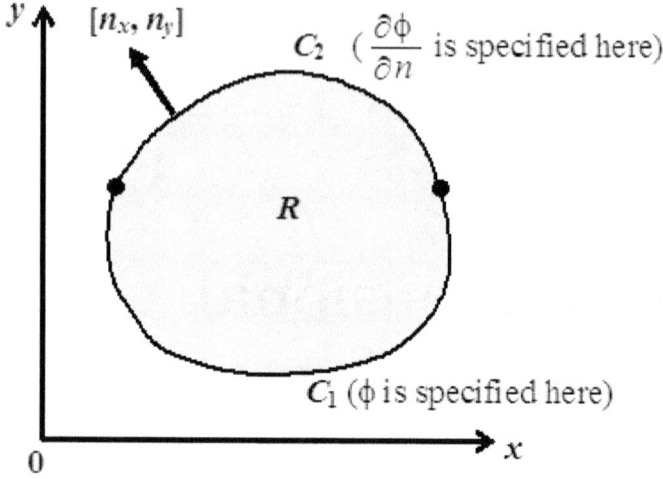

Figure 1.1

The normal derivative $\partial\phi/\partial n$ in Eq. (1.2) is defined by

$$\frac{\partial\phi}{\partial n} = n_x\frac{\partial\phi}{\partial x} + n_y\frac{\partial\phi}{\partial y}, \tag{1.3}$$

where n_x and n_y are respectively the x and y components of a unit normal vector to the curve C. Here the unit normal vector $[n_x, n_y]$ on C is taken to be pointing away from the region R. Note that the normal vector may vary from point to point on C. Thus, $[n_x, n_y]$ is a function of x and y.

The boundary conditions given in Eq. (1.2) are assumed to be properly posed so that the boundary value problem has a unique solution, that is, it is assumed that one can always find a function $\phi(x, y)$ satisfying Eqs. (1.1)-(1.2) and that there is only one such function.

For a particular example of practical situations involving the boundary value problem above, one may mention the classical heat conduction problem where ϕ denotes the steady-state temperature in an isotropic solid. Eq. (1.1) is then the temperature governing equation derived, under certain assumptions, from the law of conservation of heat energy together with the

Fourier's heat flux model. The heat flux out of the region R across the boundary C is given by $-\kappa \partial\phi/\partial n$, where κ is the thermal heat conductivity of the solid. Thus, the boundary conditions in Eq. (1.2) imply that at each and every given point on C either the temperature or the heat flux (but not both) is known. To determine the temperature field in the solid, one has to solve Eq. (1.1) in R to find the solution that satisfies the prescribed boundary conditions on C.

In general, it is difficult (if not impossible) to solve exactly the boundary value problem defined by Eqs. (1.1)-(1.2). The mathematical complexity involved depends on the geometrical shape of the region R and the boundary conditions given in Eq. (1.2). Exact solutions can only be found for relatively simple geometries of R (such as a square region) together with particular boundary conditions. For more complicated geometries or general boundary conditions, one may have to resort to numerical (approximate) techniques for solving Eqs. (1.1)-(1.2).

This chapter introduces a boundary element method for the numerical solution of the interior boundary value problem defined by Eqs. (1.1)-(1.2). We show how a boundary integral solution can be derived for Eq. (1.1) and applied to obtain a simple boundary element procedure for approximately solving the boundary value problem under consideration. The implementation of the numerical procedure on the computer, achieved through coding in FORTRAN 77, is discussed in detail.

1.2   Fundamental Solution

If we use polar coordinates r and θ centered about $(0, 0)$, as defined by $x = r \cos\theta$ and $y = r \sin\theta$, and introduce $\psi(r, \theta) = \phi(r \cos\theta, r \sin\theta)$, we can rewrite Eq. (1.1) as

$$\frac{1}{r} \frac{\partial}{\partial r}\left(r \frac{\partial \psi}{\partial r}\right) + \frac{1}{r^2} \frac{\partial^2 \psi}{\partial \theta^2} = 0. \tag{1.4}$$

For the case in which ψ is independent of θ, that is, if ψ is a function of r alone, Eq. (1.4) reduces to the ordinary

9

differential equation

$$\frac{d}{dr}(r\frac{d}{dr}[\psi(r)]) = 0 \text{ for } r \neq 0. \tag{1.5}$$

The ordinary differential equation in Eq. (1.5) can be easily integrated twice to yield the general solution

$$\psi(r) = A\ln(r) + B, \tag{1.6}$$

where A and B are arbitrary constants.

From (1.6), it is obvious that the two-dimensional Laplace's equation in Eq. (1.1) admits a class of particular solutions given by

$$\phi(x,y) = A\ln\sqrt{x^2 + y^2} + B \quad \text{for } (x,y) \neq (0,0). \tag{1.7}$$

If we choose the constants A and B in (1.7) to be $1/(2\pi)$ and 0 respectively and shift the center of the polar coordinates from $(0,0)$ to the general point (ξ, η), a particular solution of Eq. (1.1) is

$$\phi(x,y) = \frac{1}{2\pi}\ln\sqrt{(x-\xi)^2 + (y-\eta)^2} \quad \text{for } (x,y) \neq (\xi, \eta). \tag{1.8}$$

As we shall see, the particular solution in Eq. (1.8) plays an important role in the development of boundary element methods for the numerical solution of the interior boundary value problem defined by Eqs. (1.1)-(1.2). We specially denote this particular solution using the symbol $\Phi(x,y;\xi,\eta)$, that is, we write

$$\Phi(x,y;\xi,\eta) = \frac{1}{4\pi}\ln[(x-\xi)^2 + (y-\eta)^2]. \tag{1.9}$$

We refer to $\Phi(x,y;\xi,\eta)$ in Eq. (1.9) as the fundamental solution of the two-dimensional Laplace's equation. Note that $\Phi(x,y;\xi,\eta)$ satisfies Eq. (1.1) everywhere except at (ξ,η) where it is not well defined.

1.3 Reciprocal Relation

If ϕ_1 and ϕ_2 are any two solutions of Eq. (1.1) in the region R bounded by the simple closed curve C then it can be shown that

$$\int_C (\phi_2 \frac{\partial \phi_1}{\partial n} - \phi_1 \frac{\partial \phi_2}{\partial n}) ds(x, y) = 0. \qquad (1.10)$$

Eq. (1.10) provides a reciprocal relation between any two solutions of the Laplace's equation in the region R bounded by the curve C. It may be derived from the two-dimensional version of the Gauss-Ostrogradskii (divergence) theorem as explained below.

According to the divergence theorem, if $\underline{\mathbf{F}} = u(x, y)\underline{\mathbf{i}} + v(x, y)\underline{\mathbf{j}}$ is a well defined vector function such that $\underline{\nabla} \cdot \underline{\mathbf{F}} = \partial u / \partial x + \partial v / \partial y$ exists in the region R bounded by the simple closed curve C then

$$\int_C \underline{\mathbf{F}} \cdot \underline{\mathbf{n}}\, ds(x, y) = \iint_R \underline{\nabla} \cdot \underline{\mathbf{F}}\, dx dy,$$

that is,

$$\int_C [u n_x + v n_y] ds(x, y) = \iint_R [\frac{\partial u}{\partial x} + \frac{\partial v}{\partial y}] dx dy,$$

where $\underline{\mathbf{n}} = [n_x, n_y]$ is the unit normal vector to the curve C, pointing away from the region R.

Since ϕ_1 and ϕ_2 are solutions of Eq. (1.1), we may write

$$\frac{\partial^2 \phi_1}{\partial x^2} + \frac{\partial^2 \phi_1}{\partial y^2} = 0,$$

$$\frac{\partial^2 \phi_2}{\partial x^2} + \frac{\partial^2 \phi_2}{\partial y^2} = 0.$$

If we multiply the first equation by ϕ_2 and the second one by ϕ_1 and take the difference of the resulting equations, we

11

obtain

$$\frac{\partial}{\partial x}(\phi_2\frac{\partial\phi_1}{\partial x} - \phi_1\frac{\partial\phi_2}{\partial x}) + \frac{\partial}{\partial y}(\phi_2\frac{\partial\phi_1}{\partial y} - \phi_1\frac{\partial\phi_2}{\partial y}) = 0,$$

which can be integrated over R to give

$$\iint\limits_{R} [\frac{\partial}{\partial x}(\phi_2\frac{\partial\phi_1}{\partial x} - \phi_1\frac{\partial\phi_2}{\partial x}) + \frac{\partial}{\partial y}(\phi_2\frac{\partial\phi_1}{\partial y} - \phi_1\frac{\partial\phi_2}{\partial y})]dxdy = 0.$$

Application of the divergence theorem to convert the double integral over R into a line integral over C yields

$$\int\limits_{C} [(\phi_2\frac{\partial\phi_1}{\partial x} - \phi_1\frac{\partial\phi_2}{\partial x})n_x + (\phi_2\frac{\partial\phi_1}{\partial y} - \phi_1\frac{\partial\phi_2}{\partial y})n_y]ds(x,y) = 0$$

which is essentially Eq. (1.10).

Together with the fundamental solution given by Eq. (1.9), the reciprocal relation in Eq. (1.10) can be used to derive a useful boundary integral solution for the two-dimensional Laplace's equation.

1.4   Boundary Integral Solution

Let us take $\phi_1 = \Phi(x,y;\xi,\eta)$ (the fundamental solution as defined in Eq. (1.9)) and $\phi_2 = \phi$, where ϕ is the required solution of the interior boundary value problem defined by Eqs. (1.1)-(1.2).

Since $\Phi(x,y;\xi,\eta)$ is not well defined at the point (ξ,η), the reciprocal relation in Eq. (1.10) is valid for $\phi_1 = \Phi(x,y;\xi,\eta)$ and $\phi_2 = \phi$ only if (ξ,η) does not lie in the region $R\cup C$. Thus,

$$\int\limits_{C} [\phi(x,y)\frac{\partial}{\partial n}(\Phi(x,y;\xi,\eta))$$

$$-\Phi(x,y;\xi,\eta)\frac{\partial}{\partial n}(\phi(x,y))]ds(x,y)$$
$$= \quad 0 \text{ for } (\xi,\eta) \notin R\cup C. \qquad (1.11)$$

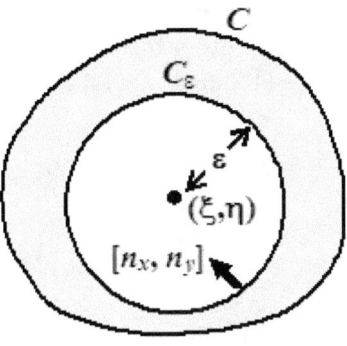

Figure 1.2

A more interesting and useful integral equation than Eq. (1.11) can be derived from Eq. (1.10) if we take the point (ξ, η) to lie in the region $R \cup C$.

For the case in which (ξ, η) lies in the interior of R, Eq. (1.10) is valid if we replace C by $C \cup C_\varepsilon$, where C_ε is a circle of center (ξ, η) and radius ε as shown in Figure 1.2[1]. This is because $\Phi(x, y; \xi, \eta)$ and its first order partial derivatives (with respect to x or y) are well defined in the region between C and C_ε. Thus, for C and C_ε in Figure 1.2, we can write

$$
\int_C [\phi(x,y) \frac{\partial}{\partial n}(\Phi(x,y;\xi,\eta))
$$

$$
- \Phi(x,y;\xi,\eta) \frac{\partial}{\partial n}(\phi(x,y))] ds(x,y)
$$

$$
+ \int_{C_\varepsilon} [\phi(x,y) \frac{\partial}{\partial n}(\Phi(x,y;\xi,\eta))
$$

$$
- \Phi(x,y;\xi,\eta) \frac{\partial}{\partial n}(\phi(x,y))] ds(x,y)
$$

$$
= 0. \tag{1.12}
$$

---

[1] The divergence theorem is not only applicable for simply connected regions but also for multiply connected ones such as the one shown in Figure 1.2. For the region in Figure 1.2, the unit normal vector to C_ε (the inner boundary) points towards the center of the circle.

Eq. (1.12) holds for any radius $\varepsilon > 0$, so long as the circle C_ε (in Figure 1.2) lies completely inside the region bounded by C. Thus, we may let $\varepsilon \to 0^+$ in Eq. (1.12). This gives

$$\int_C [\phi(x,y)\frac{\partial}{\partial n}(\Phi(x,y;\xi,\eta))$$

$$-\Phi(x,y;\xi,\eta)\frac{\partial}{\partial n}(\phi(x,y))]ds(x,y)$$

$$= -\lim_{\varepsilon \to 0^+} \int_{C_\varepsilon} [\phi(x,y)\frac{\partial}{\partial n}(\Phi(x,y;\xi,\eta))$$

$$-\Phi(x,y;\xi,\eta)\frac{\partial}{\partial n}(\phi(x,y))]ds(x,y). \qquad (1.13)$$

Using polar coordinates r and θ centered about (ξ,η) as defined by $x - \xi = r\cos\theta$ and $y - \eta = r\sin\theta$, we may write

$$\Phi(x,y;\xi,\eta) = \frac{1}{2\pi}\ln(r),$$

$$\frac{\partial}{\partial n}[\Phi(x,y;\xi,\eta)] = n_x\frac{\partial}{\partial x}[\Phi(x,y;\xi,\eta)] + n_y\frac{\partial}{\partial y}[\Phi(x,y;\xi,\eta)]$$

$$= \frac{n_x\cos\theta + n_y\sin\theta}{2\pi r}. \qquad (1.14)$$

The Taylor's series of $\phi(x,y)$ about the point (ξ,η) is given by

$$\phi(x,y) = \sum_{m=0}^{\infty}\sum_{k=0}^{m}(\frac{\partial^m\phi}{\partial x^k\partial y^{m-k}})\Big|_{(x,y)=(\xi,\eta)}\frac{(x-\xi)^k(y-\eta)^{m-k}}{k!(m-k)!}.$$

On the circle C_ε, $r = \varepsilon$. Thus,

$$\phi(x,y) = \sum_{m=0}^{\infty}\sum_{k=0}^{m}(\frac{\partial^m}{\partial x^k\partial y^{m-k}}[\phi(x,y)])\Big|_{(x,y)=(\xi,\eta)}$$

$$\times\frac{\varepsilon^m\cos^k\theta\sin^{m-k}\theta}{k!(m-k)!} \quad \text{for } (x,y) \in C_\varepsilon. \qquad (1.15)$$

14

Similarly, we may write

$$\frac{\partial}{\partial n}[\phi(x,y)] = \sum_{m=0}^{\infty}\sum_{k=0}^{m} \left(\frac{\partial^m}{\partial x^k \partial y^{m-k}}\left\{\frac{\partial}{\partial n}[\phi(x,y)]\right\}\right)\bigg|_{(x,y)=(\xi,\eta)}$$
$$\times \frac{\varepsilon^m \cos^k\theta \sin^{m-k}\theta}{k!(m-k)!} \quad \text{for } (x,y) \in C_{\varepsilon}.$$
(1.16)

Using Eqs. (1.14), (1.15) and (1.16) and writing $ds(x,y) = \varepsilon d\theta$ with θ ranging from 0 to 2π, we may now attempt to evaluate the limit on the right hand side of Eq. (1.13). On C_{ε}, the normal vector $[n_x, n_y]$ is given by $[-\cos\theta, -\sin\theta]$. Thus,

$$\int_{C_{\varepsilon}} \phi(x,y)\frac{\partial}{\partial n}[\Phi(x,y;\xi,\eta)]ds(x,y)$$

$$= \quad -\frac{1}{2\pi}\phi(\xi,\eta)\int_{0}^{2\pi} d\theta$$

$$-\frac{1}{2\pi}\sum_{m=1}^{\infty}\sum_{k=0}^{m}\frac{\varepsilon^m}{k!(m-k)!}\left(\frac{\partial^m\phi}{\partial x^k\partial y^{m-k}}\right)\bigg|_{(x,y)=(\xi,\eta)}$$

$$\times \int_{0}^{2\pi}\cos^k\theta\sin^{m-k}\theta d\theta$$

$$\rightarrow \quad -\phi(\xi,\eta) \quad \text{as } \varepsilon \rightarrow 0^+, \qquad (1.17)$$

and

$$\int_{C_{\varepsilon}} \Phi(x,y;\xi,\eta)\frac{\partial}{\partial n}[\phi(x,y)]ds(x,y)$$

$$= \quad \frac{1}{2\pi}\sum_{m=0}^{\infty}\sum_{k=0}^{m}\left(\frac{\partial^m}{\partial x^k\partial y^{m-k}}\left(\frac{\partial}{\partial n}[\phi(x,y)]\right)\right)\bigg|_{(x,y)=(\xi,\eta)}$$

$$\times \frac{\varepsilon^{m+1}\ln(\varepsilon)}{k!(m-k)!}\int_{0}^{2\pi}\cos^k\theta\sin^{m-k}\theta d\theta$$

$$\rightarrow \quad 0 \quad \text{as } \varepsilon \rightarrow 0^+, \qquad (1.18)$$

since $\varepsilon^{m+1}\ln(\varepsilon) \to 0$ as $\varepsilon \to 0^+$ for $m = 0, 1, 2, \cdots$.

Consequently, as $\varepsilon \to 0^+$, Eq. (1.13) yields

$$
\phi(\xi,\eta) = \int_C [\phi(x,y)\frac{\partial}{\partial n}(\Phi(x,y;\xi,\eta))
$$

$$
-\Phi(x,y;\xi,\eta)\frac{\partial}{\partial n}(\phi(x,y))]ds(x,y)
$$

$$
\text{for } (\xi,\eta) \in R. \tag{1.19}
$$

Together with Eq. (1.9), Eq. (1.19) provides us with a boundary integral solution for the two-dimensional Laplace's equation. If both ϕ and $\partial\phi/\partial n$ are known at all points on C, the line integral in Eq. (1.19) can be evaluated (at least in theory) to calculate ϕ at any point (ξ,η) in the interior of R. From the boundary conditions (1.2), however, at any given point on C, either ϕ or $\partial\phi/\partial n$, not both, is known.

To solve the interior boundary value problem, we must find the unknown ϕ and $\partial\phi/\partial n$ on C_2 and C_1 respectively. As we shall see later on, this may be done through manipulation of data on the boundary C only, if we can derive a boundary integral formula for $\phi(\xi,\eta)$, similar to the one in Eq. (1.19), for a general point (ξ,η) that lies on C.

For the case in which the point (ξ,η) lies on C, Eq. (1.10) holds if we replace the curve C by $D \cup D_\varepsilon$, where the curves D and D_ε are as shown in Figure 1.3. (If C_ε is the circle of center (ξ,η) and radius ε, then D is the part of C that lies outside C_ε and D_ε is the part of C_ε that is inside R.) Thus,

$$
\int_D [\phi(x,y)\frac{\partial}{\partial n}(\Phi(x,y;\xi,\eta))
$$

$$
-\Phi(x,y;\xi,\eta)\frac{\partial}{\partial n}(\phi(x,y))]ds(x,y)
$$

$$
= -\int_{D_\varepsilon} [\phi(x,y)\frac{\partial}{\partial n}(\Phi(x,y;\xi,\eta))
$$

$$
-\Phi(x,y;\xi,\eta)\frac{\partial}{\partial n}(\phi(x,y))]ds(x,y). \tag{1.20}
$$

16

Let us examine what happens to Eq. (1.20) when we let $\varepsilon \to 0^+$.

As $\varepsilon \to 0^+$, the curve D tends to C. Thus, we may write

$$\int_C [\phi(x,y)\frac{\partial}{\partial n}(\Phi(x,y;\xi,\eta))$$

$$-\Phi(x,y;\xi,\eta)\frac{\partial}{\partial n}(\phi(x,y))]ds(x,y)$$

$$= -\lim_{\varepsilon \to 0^+} \int_{D_\varepsilon} [\phi(x,y)\frac{\partial}{\partial n}(\Phi(x,y;\xi,\eta))$$

$$-\Phi(x,y;\xi,\eta)\frac{\partial}{\partial n}(\phi(x,y))]ds(x,y). \qquad (1.21)$$

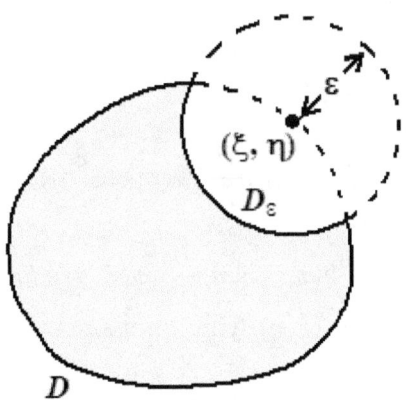

Figure 1.3

Note that, unlike in Eq. (1.13), the line integral over C in Eq. (1.21) is improper as its integrand is not well defined at (ξ,η) which lies on C. Strictly speaking, the line integration should be over the curve C without an infinitesimal segment that contains the point (ξ,η), that is, the line integral over C in Eq. (1.21) has to be interpreted in the Cauchy principal sense if (ξ,η) lies on C.

To evaluate the limit on the right hand side of Eq. (1.21), we need to know what happens to D_ε when we let $\varepsilon \to 0^+$. Now if (ξ, η) lies on a smooth part of C (not at where the gradient of the curve changes abruptly, that is, not at a corner point, if there is any), one can intuitively see that the part of C inside C_ε approaches an infinitesimal straight line as $\varepsilon \to 0^+$. Thus, we expect D_ε to tend to a semi-circle as $\varepsilon \to 0^+$, if (ξ, η) lies on a smooth part of C. It follows that in attempting to evaluate the limit on the right hand side of Eq. (1.21) we have to integrate over only half a circle (instead of a full circle as in the case of Eq. (1.13)).

Modifying Eqs. (1.17) and (1.18), we obtain

$$\lim_{\varepsilon \to 0^+} \int_{D_\varepsilon} \phi(x,y) \frac{\partial}{\partial n} [\Phi(x,y;\xi,\eta)] ds(x,y) = -\frac{1}{2}\phi(\xi,\eta),$$

$$\lim_{\varepsilon \to 0^+} \int_{D_\varepsilon} \Phi(x,y;\xi,\eta) \frac{\partial}{\partial n} [\phi(x,y)] ds(x,y) = 0.$$

Hence Eq. (1.21) gives

$$\frac{1}{2}\phi(\xi,\eta) = \int_C [\phi(x,y) \frac{\partial}{\partial n}(\Phi(x,y;\xi,\eta))$$

$$-\Phi(x,y;\xi,\eta) \frac{\partial}{\partial n}(\phi(x,y))] ds(x,y)$$

$$\text{for } (\xi,\eta) \text{ lying on a smooth part of } C.$$

$$(1.22)$$

Together with the boundary conditions in Eq. (1.2), Eq. (1.22) may be applied to obtain a numerical procedure for determining the unknown ϕ and/or $\partial\phi/\partial n$ on the boundary C. Once ϕ and $\partial\phi/\partial n$ are known at all points on C, the solution of the interior boundary value problem defined by Eqs. (1.1)-(1.2) is given by Eq. (1.19) at any point (ξ,η) inside R. More details are given in Section 1.5 below.

For convenience, we may write Eqs. (1.11), (1.19) and (1.22) as a single equation given by

$$\lambda(\xi,\eta)\phi(\xi,\eta) \;=\; \int\limits_{C} [\phi(x,y)\frac{\partial}{\partial n}(\Phi(x,y;\xi,\eta))$$

$$-\Phi(x,y;\xi,\eta)\frac{\partial}{\partial n}(\phi(x,y))]ds(x,y),$$

$$(1.23)$$

if we define

$$\lambda(\xi,\eta) = \left\{ \begin{array}{ll} 0 & \text{if } (\xi,\eta) \notin R \cup C, \\ 1/2 & \text{if } (\xi,\eta) \text{ lies on a smooth part of } C, \\ 1 & \text{if } (\xi,\eta) \in R. \end{array} \right.$$

$$(1.24)$$

1.5 Boundary Element Solution with Constant Elements

We now show how Eq. (1.23) may be applied to obtain a simple boundary element procedure for solving numerically the interior boundary value problem defined by Eqs. (1.1)-(1.2).

The boundary C is discretized into N very small straight line segments $C^{(1)}$, $C^{(2)}$, $\cdots$, $C^{(N-1)}$ and $C^{(N)}$, that is,

$$C \simeq C^{(1)} \cup C^{(2)} \cup \cdots \cup C^{(N-1)} \cup C^{(N)}. \qquad (1.25)$$

The sides or the boundary elements $C^{(1)}$, $C^{(2)}$, $\cdots$, $C^{(N-1)}$ and $C^{(N)}$ are constructed as follows. We put N well spaced out points given by $(x^{(1)},y^{(1)})$, $(x^{(2)},y^{(2)})$, $\cdots$, $(x^{(N-1)},y^{(N-1)})$ and $(x^{(N)},y^{(N)})$ on C, in the order given, following the counter clockwise direction. Defining $(x^{(N+1)},y^{(N+1)}) = (x^{(1)},y^{(1)})$, we take $C^{(k)}$ to be the boundary element from $(x^{(k)},y^{(k)})$ to $(x^{(k+1)},y^{(k+1)})$ for $k = 1, 2, \cdots , N$.

As an example, in Figure 1.4, the boundary $C = C_1 \cup C_2$ in Figure 1.1 is approximated using 5 boundary elements denoted by $C^{(1)}$, $C^{(2)}$, $C^{(3)}$, $C^{(4)}$ and $C^{(5)}$.

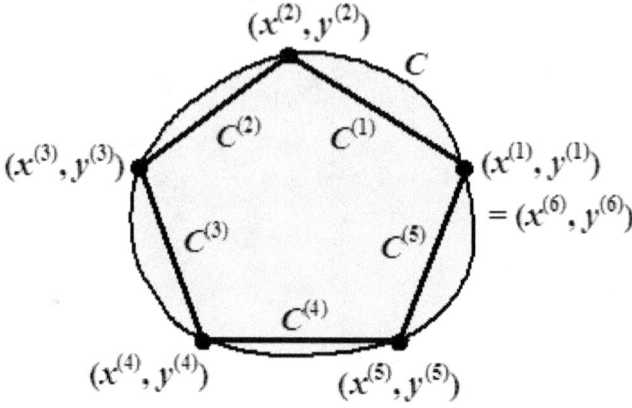

Figure 1.4

For a simple approximation of ϕ and $\partial\phi/\partial n$ on the boundary C, we assume that these functions are constants over each of the boundary elements. Specifically, we make the approximation:

$$\phi \simeq \overline{\phi}^{(k)} \quad \text{and} \quad \frac{\partial\phi}{\partial n} = \overline{p}^{(k)} \quad \text{for } (x, y) \in C^{(k)} \ (k = 1, 2, \cdots, N),$$
$$(1.26)$$

where $\overline{\phi}^{(k)}$ and $\overline{p}^{(k)}$ are respectively the values of ϕ and $\partial\phi/\partial n$ at the midpoint of $C^{(k)}$.

With Eqs. (1.25) and (1.26), we find that Eq. (1.23) can be approximately written as

$$\lambda(\xi, \eta)\phi(\xi, \eta) = \sum_{k=1}^{N}\{\overline{\phi}^{(k)}\mathcal{F}_2^{(k)}(\xi, \eta) - \overline{p}^{(k)}\mathcal{F}_1^{(k)}(\xi, \eta)\}, \quad (1.27)$$

where

$$\mathcal{F}_1^{(k)}(\xi, \eta) = \int_{C^{(k)}} \Phi(x, y; \xi, \eta)ds(x, y),$$

$$\mathcal{F}_2^{(k)}(\xi, \eta) = \int_{C^{(k)}} \frac{\partial}{\partial n}[\Phi(x, y; \xi, \eta)]ds(x, y). \quad (1.28)$$

20

For a given k, either $\overline{\phi}^{(k)}$ or $\overline{p}^{(k)}$ (not both) is known from the boundary conditions in Eq. (1.2). Thus, there are N unknown constants on the right hand side of Eq. (1.27). To determine their values, we have to generate N equations containing the unknowns.

If we let (ξ, η) in Eq. (1.27) be given in turn by the midpoints of $C^{(1)}, C^{(2)}, \cdots, C^{(N-1)}$ and $C^{(N)}$, we obtain

$$\frac{1}{2}\overline{\phi}^{(m)} = \sum_{k=1}^{N}\{\overline{\phi}^{(k)}\mathcal{F}_2^{(k)}(\overline{x}^{(m)}, \overline{y}^{(m)}) - \overline{p}^{(k)}\mathcal{F}_1^{(k)}(\overline{x}^{(m)}, \overline{y}^{(m)})\}$$

$$\text{for } m = 1, 2, \cdots, N, \qquad (1.29)$$

where $(\overline{x}^{(m)}, \overline{y}^{(m)})$ is the midpoint of $C^{(m)}$.

In the derivation of Eq. (1.29), we take $\lambda(\overline{x}^{(m)}, \overline{y}^{(m)}) = 1/2$, since $(\overline{x}^{(m)}, \overline{y}^{(m)})$ being the midpoint of $C^{(m)}$ lies on a smooth part of the approximate boundary $C^{(1)} \cup C^{(2)} \cup \cdots \cup C^{(N-1)} \cup C^{(N)}$.

Eq. (1.29) constitutes a system of N linear algebraic equations containing the N unknowns on the right hand side of Eq. (1.27). We may rewrite it as

$$\sum_{k=1}^{N} a^{(mk)} z^{(k)} = \sum_{k=1}^{N} b^{(mk)} \text{ for } m = 1, 2, \cdots, N, \qquad (1.30)$$

where $a^{(mk)}$, $b^{(mk)}$ and $z^{(k)}$ are defined by

$$a^{(mk)} = \begin{cases} -\mathcal{F}_1^{(k)}(\overline{x}^{(m)}, \overline{y}^{(m)}) \text{ if } \phi \text{ is specified over } C^{(k)}, \\ \mathcal{F}_2^{(k)}(\overline{x}^{(m)}, \overline{y}^{(m)}) - \frac{1}{2}\delta^{(mk)} \text{ if } \partial\phi/\partial n \text{ is} \\ \qquad\qquad\qquad\qquad\qquad \text{specified over } C^{(k)}, \end{cases}$$

$$b^{(mk)} = \begin{cases} \overline{\phi}^{(k)}(-\mathcal{F}_2^{(k)}(\overline{x}^{(m)}, \overline{y}^{(m)}) + \frac{1}{2}\delta^{(mk)}) \\ \qquad\qquad\qquad \text{if } \phi \text{ is specified over } C^{(k)}, \\ \overline{p}^{(k)}\mathcal{F}_1^{(k)}(\overline{x}^{(m)}, \overline{y}^{(m)}) \text{ if } \partial\phi/\partial n \text{ is specified} \\ \qquad\qquad\qquad\qquad\qquad \text{over } C^{(k)}, \end{cases}$$

$$\delta^{(mk)} = \begin{cases} 0 & \text{if } m \neq k, \\ 1 & \text{if } m = k, \end{cases}$$

$$z^{(k)} = \begin{cases} \overline{p}^{(k)} & \text{if } \phi \text{ is specified over } C^{(k)}, \\ \overline{\phi}^{(k)} & \text{if } \partial\phi/\partial n \text{ is specified over } C^{(k)}. \end{cases} \qquad (1.31)$$

Note that $z^{(1)}$, $z^{(2)}$, $\cdots$, $z^{(N-1)}$ and $z^{(N)}$ are the N unknown constants on the right hand side of Eq. (1.27), while $a^{(mk)}$ and $b^{(mk)}$ are known coefficients.

Once Eq. (1.30) is solved for the unknowns $z^{(1)}$, $z^{(2)}$, $\cdots$, $z^{(N-1)}$ and $z^{(N)}$, the values of ϕ and $\partial\phi/\partial n$ over the element $C^{(k)}$, as given by $\overline{\phi}^{(k)}$ and $\overline{p}^{(k)}$ respectively, are known for $k = 1$, 2, $\cdots$, N. Eq. (1.27) with $\lambda(\xi, \eta) = 1$ then provides us with an explicit formula for computing ϕ in the interior of R, that is,

$$\phi(\xi, \eta) \simeq \sum_{k=1}^{N} \{\overline{\phi}^{(k)} \mathcal{F}_2^{(k)}(\xi, \eta) - \overline{p}^{(k)} \mathcal{F}_1^{(k)}(\xi, \eta)\} \quad \text{for } (\xi, \eta) \in R.$$

(1.32)

To summarize, a boundary element solution of the interior boundary value problem defined by Eqs. (1.1)-(1.2) is given by Eq. (1.32) together with Eqs. (1.28), (1.30) and (1.31). Because of the approximation in Eqs. (1.25) and (1.26), the solution is said to be obtained using constant elements. Analytical formulae for calculating $\mathcal{F}_1^{(k)}(\xi, \eta)$ and $\mathcal{F}_2^{(k)}(\xi, \eta)$ in Eq. (1.28) are given in Eqs. (1.37), (1.38), (1.40) and (1.41) (together with Eq. (1.35)) in the section below.

1.6 Formulae for Integrals of Constant Elements

The boundary element solution above requires the evaluation of $\mathcal{F}_1^{(k)}(\xi, \eta)$ and $\mathcal{F}_2^{(k)}(\xi, \eta)$. These functions are defined in terms of line integrals over $C^{(k)}$ as given in Eq. (1.28). The line integrals can be worked out analytically as follows.

Points on the element $C^{(k)}$ may be described using the parametric equations

$$\left. \begin{array}{l} x = x^{(k)} - t\ell^{(k)} n_y^{(k)} \\ y = y^{(k)} + t\ell^{(k)} n_x^{(k)} \end{array} \right\} \quad \text{from } t = 0 \text{ to } t = 1, \qquad (1.33)$$

where $\ell^{(k)}$ is the length of $C^{(k)}$ and $[n_x^{(k)}, n_y^{(k)}] = [y^{(k+1)} - y^{(k)}, x^{(k)} - x^{(k+1)}]/\ell^{(k)}$ is the unit normal vector to $C^{(k)}$ pointing away from R.

For $(x, y) \in C^{(k)}$, we find that $ds(x,y) = \sqrt{(dx)^2 + (dy)^2} = \ell^{(k)}dt$ and

$$(x - \xi)^2 + (y - \eta)^2 = A^{(k)}t^2 + B^{(k)}(\xi, \eta)t + E^{(k)}(\xi, \eta), \quad (1.34)$$

where

$$A^{(k)} = [\ell^{(k)}]^2,$$
$$B^{(k)}(\xi, \eta) = [-n_y^{(k)}(x^{(k)} - \xi) + (y^{(k)} - \eta)n_x^{(k)}](2\ell^{(k)}),$$
$$E^{(k)}(\xi, \eta) = (x^{(k)} - \xi)^2 + (y^{(k)} - \eta)^2. \quad (1.35)$$

For any point (ξ, η), the parameters in Eq. (1.35) satisfy $4A^{(k)}E^{(k)}(\xi, \eta) - [B^{(k)}(\xi, \eta)]^2 \geq 0$. To see why this is true, consider the straight line defined by the parametric equations $x = x^{(k)} - t\ell^{(k)}n_y^{(k)}$ and $y = y^{(k)} + t\ell^{(k)}n_x^{(k)}$ for $-\infty < t < \infty$. Note that $C^{(k)}$ is a subset of this straight line (given by the parametric equations from $t = 0$ to $t = 1$). Eq. (1.34) also holds for any point (x, y) lying on the extended line. If (ξ, η) does not lie on the line then $A^{(k)}t^2 + B^{(k)}(\xi, \eta)t + E^{(k)}(\xi, \eta) > 0$ for all real values of t (that is, for all points (x, y) on the line) and hence $4A^{(k)}E^{(k)}(\xi, \eta) - [B^{(k)}(\xi, \eta)]^2 > 0$. On the other hand, if (ξ, η) is on the line, we can find exactly one point (x, y) such that $A^{(k)}t^2 + B^{(k)}(\xi, \eta)t + E^{(k)}(\xi, \eta) = 0$. As each point (x, y) on the line is given by a unique value of t, we conclude that $4A^{(k)}E^{(k)}(\xi, \eta) - [B^{(k)}(\xi, \eta)]^2 = 0$ for (ξ, η) lying on the line.

From Eqs. (1.28), (1.33) and (1.34), $\mathcal{F}_1^{(k)}(\xi, \eta)$ and $\mathcal{F}_2^{(k)}(\xi, \eta)$ may be written as

$$\mathcal{F}_1^{(k)}(\xi, \eta) = \frac{\ell^{(k)}}{4\pi} \int_0^1 \ln[A^{(k)}t^2 + B^{(k)}(\xi, \eta)t + E^{(k)}(\xi, \eta)]dt,$$

$$\mathcal{F}_2^{(k)}(\xi, \eta) = \frac{\ell^{(k)}}{2\pi} \int_0^1 \frac{n_x^{(k)}(x^{(k)} - \xi) + n_y^{(k)}(y^{(k)} - \eta)}{A^{(k)}t^2 + B^{(k)}(\xi, \eta)t + E^{(k)}(\xi, \eta)}dt.$$

$$(1.36)$$

The second integral in Eq. (1.36) is the easiest one to work out for the case in which $4A^{(k)}E^{(k)}(\xi, \eta) - [B^{(k)}(\xi, \eta)]^2 = 0$. For this case, the point (ξ, η) lies on the straight line of which the element $C^{(k)}$ is a subset. Thus, the vector $[x^{(k)} - \xi, y^{(k)} - \eta]$ is perpendicular to $[n_x^{(k)}, n_y^{(k)}]$, that is, $n_x^{(k)}(x^{(k)} - \xi) + n_y^{(k)}(y^{(k)} - \eta) = 0$, and we obtain

$$\mathcal{F}_2^{(k)}(\xi, \eta) = 0 \quad \text{for } 4A^{(k)}E^{(k)}(\xi, \eta) - [B^{(k)}(\xi, \eta)]^2 = 0. \quad (1.37)$$

From the integration formula

$$\int \frac{dt}{at^2 + bt + c} = \frac{2}{\sqrt{4ac - b^2}} \arctan\left(\frac{2at + b}{\sqrt{4ac - b^2}}\right) + \text{constant}$$

for real constants a, b and c

such that $4ac - b^2 > 0$,

we find that

$$\mathcal{F}_2^{(k)}(\xi, \eta) = \frac{\ell^{(k)}[n_x^{(k)}(x^{(k)} - \xi) + n_y^{(k)}(y^{(k)} - \eta)]}{\pi\sqrt{4A^{(k)}E^{(k)}(\xi, \eta) - [B^{(k)}(\xi, \eta)]^2}}$$

$$\times[\arctan\left(\frac{2A^{(k)} + B^{(k)}(\xi, \eta)}{\sqrt{4A^{(k)}E^{(k)}(\xi, \eta) - [B^{(k)}(\xi, \eta)]^2}}\right)$$

$$- \arctan\left(\frac{B^{(k)}(\xi, \eta)}{\sqrt{4A^{(k)}E^{(k)}(\xi, \eta) - [B^{(k)}(\xi, \eta)]^2}}\right)]$$

$$\text{for } 4A^{(k)}E^{(k)}(\xi, \eta) - [B^{(k)}(\xi, \eta)]^2 > 0. \quad (1.38)$$

If $4A^{(k)}E^{(k)}(\xi, \eta) - [B^{(k)}(\xi, \eta)]^2 = 0$, we may write

$$A^{(k)}t^2 + B^{(k)}(\xi, \eta)t + E^{(k)}(\xi, \eta) = A^{(k)}\left(t + \frac{B^{(k)}(\xi, \eta)}{2A^{(k)}}\right)^2.$$

Thus,

$$\mathcal{F}_1^{(k)}(\xi, \eta) = \frac{\ell^{(k)}}{4\pi} \int_0^1 \ln\left[A^{(k)}\left(t + \frac{B^{(k)}(\xi, \eta)}{2A^{(k)}}\right)^2\right]dt$$

$$\text{for } 4A^{(k)}E^{(k)}(\xi, \eta) - [B^{(k)}(\xi, \eta)]^2 = 0.$$

$$(1.39)$$

Now if (ξ, η) lies on a smooth part of $C^{(k)}$, the integral in Eq. (1.39) is improper, as its integrand is not well defined at the point $t = t_0 \equiv -B^{(k)}(\xi, \eta)/(2A^{(k)}) \in (0,1)$. Strictly speaking, the integral should then be interpreted in the Cauchy principal sense, that is, to evaluate it, we have to integrate over $[0, t_0 - \varepsilon] \cup [t_0 + \varepsilon, 1]$ instead of $[0, 1]$ and then let $\varepsilon \to 0$ to obtain its value. In this case, however, it turns out that the limits of integration $t = t_0 - \varepsilon$ and $t = t_0 + \varepsilon$ eventually do not contribute anything to the integral. Thus, for $4A^{(k)} E^{(k)}(\xi, \eta) - [B^{(k)}(\xi, \eta)]^2 = 0$, the final analytical formula for $\mathcal{F}_1^{(k)}(\xi, \eta)$ is the same irrespective of whether (ξ, η) lies on $C^{(k)}$ or not. If (ξ, η) lies on $C^{(k)}$, we may ignore the singular behaviour of the integrand and apply the fundamental theorem of integral calculus as usual to evaluate the definite integral in Eq. (1.39) directly over $[0, 1]$.

The integration required in Eq. (1.39) can be easily done to give

$$
\begin{aligned}
\mathcal{F}_1^{(k)}(\xi, \eta) = \ & \frac{\ell^{(k)}}{2\pi} \{ \ln(\ell^{(k)}) \\
& + (1 + \frac{B^{(k)}(\xi, \eta)}{2A^{(k)}}) \ln |1 + \frac{B^{(k)}(\xi, \eta)}{2A^{(k)}}| \\
& - \frac{B^{(k)}(\xi, \eta)}{2A^{(k)}} \ln |\frac{B^{(k)}(\xi, \eta)}{2A^{(k)}}| - 1 \} \\
& \text{for } 4A^{(k)} E^{(k)}(\xi, \eta) - [B^{(k)}(\xi, \eta)]^2 = 0.
\end{aligned}
$$

$$(1.40)$$

Using

$$
\begin{aligned}
& \int \ln(at^2 + bt + c)\,dt \\
= \ & t[\ln(a) - 2] + (t + \frac{b}{2a}) \ln[t^2 + \frac{b}{a}t + \frac{c}{a}] \\
& + \frac{1}{a}\sqrt{4ac - b^2} \arctan(\frac{2at + b}{\sqrt{4ac - b^2}}) + \text{ constant} \\
& \text{for real constants } a, \ b \text{ and } c \\
& \text{such that } 4ac - b^2 > 0,
\end{aligned}
$$

we obtain

$$\mathcal{F}_1^{(k)}(\xi,\eta) = \frac{\ell^{(k)}}{4\pi}\Big\{2[\ln(\ell^{(k)}) - 1] - \frac{B^{(k)}(\xi,\eta)}{2A^{(k)}}\ln\Big|\frac{E^{(k)}(\xi,\eta)}{A^{(k)}}\Big|$$

$$+(1 + \frac{B^{(k)}(\xi,\eta)}{2A^{(k)}})\ln\Big|1 + \frac{B^{(k)}(\xi,\eta)}{A^{(k)}} + \frac{E^{(k)}(\xi,\eta)}{A^{(k)}}\Big|$$

$$+\frac{\sqrt{4A^{(k)}E^{(k)}(\xi,\eta) - [B^{(k)}(\xi,\eta)]^2}}{A^{(k)}}$$

$$\times[\arctan(\frac{2A^{(k)} + B^{(k)}(\xi,\eta)}{\sqrt{4A^{(k)}E^{(k)}(\xi,\eta) - [B^{(k)}(\xi,\eta)]^2}})$$

$$-\arctan(\frac{B^{(k)}(\xi,\eta)}{\sqrt{4A^{(k)}E^{(k)}(\xi,\eta) - [B^{(k)}(\xi,\eta)]^2}})]\Big\}$$

$$\text{for } 4A^{(k)}E^{(k)}(\xi,\eta) - [B^{(k)}(\xi,\eta)]^2 > 0.$$

$$(1.41)$$

1.7   Implementation on Computer

We attempt now to develop double precision FORTRAN 77 codes which can be used to implement the boundary element procedure described in Section 1.5 on the computer. In our discussion here, syntaxes, variables and statements in FORTRAN 77 are written in typewriter fonts, for example, `xi`, `eta` and `A=L**2d0`.

One of the tasks involved is the setting up of the system of linear algebraic equations given in Eqs. (1.30) and (1.31). To do this, the functions $\mathcal{F}_1^{(k)}(\xi,\eta)$ and $\mathcal{F}_2^{(k)}(\xi,\eta)$ have to be computed using the formulae in Section 1.6. We create a subroutine called `CPF` which accepts the values of ξ, η, $x^{(k)}$, $y^{(k)}$, $n_x^{(k)}$, $n_y^{(k)}$ and $\ell^{(k)}$(stored in the real variables `xi`, `eta`, `xk`, `yk`, `nkx`, `nky` and `L`) in order to calculate and return the values of $\pi\mathcal{F}_1^{(k)}(\xi,\eta)$ and $\pi\mathcal{F}_2^{(k)}(\xi,\eta)$ (in the real variables `PF1` and `PF2`). The subroutine `CPF` is listed below.

```
subroutine CPF(xi,eta,xk,yk,nkx,nky,L,PF1,PF2)

double precision xi,eta,xk,yk,nkx,nky,L,PF1,PF2,
```

```
      & A,B,E,D,BA,EA

        A=L**2d0
        B=2d0*L*(-nky*(xk-xi)+nkx*(yk-eta))
        E=(xk-xi)**2d0+(yk-eta)**2d0
        D=dsqrt(dabs(4d0*A*E-B**2d0))
        BA=B/A
        EA=E/A

        if (D.lt.0.0000000001d0) then
        PF1=0.5d0*L*(dlog(L)
      & +(1d0+0.5d0*BA)*dlog(dabs(1d0+0.5d0*BA))
      & -0.5d0*BA*dlog(dabs(0.5d0*BA))-1d0)
        PF2=0d0
        else
        PF1=0.25d0*L*(2d0*(dlog(L)-1d0)
      & -0.5d0*BA*dlog(dabs(EA))
      & +(1d0+0.5d0*BA)*dlog(dabs(1d0+BA+EA))
      & +(D/A)*(datan((2d0*A+B)/D)-datan(B/D)))
        PF2=L*(nkx*(xk-xi)+nky*(yk-eta))/D
      & *(datan((2d0*A+B)/D)-datan(B/D))
        endif

        return
        end
```

CPF is repeatedly called in the subroutine CELAP1. CELAP1 reads in the number of boundary elements (N) in the real variable N, the midpoints $(\overline{x}^{(k)}, \overline{y}^{(k)})$ in the real arrays xm(1:N) and ym(1:N), the boundary points $(x^{(k)}, y^{(k)})$ in the real arrays xb(1:N+1) and yb(1:N+1), the normal vectors $(n_x^{(k)}, n_y^{(k)})$ in the real arrays nx(1:N) and ny(1:N), the lengths of the boundary elements in the real array lg(1:N) and the types of boundary conditions (on the boundary elements) in the integer array BCT(1:N) together with the corresponding boundary values in the real array BCV(1:N), set up and solve Eq. (1.30), and return all the values of $\overline{\phi}^{(k)}$ and $\overline{p}^{(k)}$ in the arrays phi(1:N) and

27

dphi(1:N) respectively. (More details on the arrays BCT(1:N) and BCV(1:N) will be given later on in Section 1.8.) Thus, a large part of the boundary element procedure (with constant elements) for the numerical solution of the boundary value problem is executed in CELAP1.

The subroutine CELAP1 is listed as follows.

```
    subroutine CELAP1(N,xm,ym,xb,yb,nx,ny,lg,
& BCT,BCV,phi,dphi)

    integer m,k,N,BCT(1000)

    double precision xm(1000),ym(1000),xb(1000),
& yb(1000),nx(1000),ny(1000),lg(1000),
& BCV(1000),A(1000,1000),B(1000),pi,PF1,PF2,
& del,phi(1000),dphi(1000),F1,F2,Z(1000)

    pi=4d0*datan(1d0)

    do 10 m=1,N
    B(m)=0d0
    do 5 k=1,N
    call CPF(xm(m),ym(m),xb(k),yb(k),nx(k),ny(k),
& lg(k),PF1,PF2)
    F1=PF1/pi
    F2=PF2/pi
    if (k.eq.m) then
    del=1d0
    else
    del=0d0
    endif
    if (BCT(k).eq.0) then
    A(m,k)=-F1
    B(m)=B(m)+BCV(k)*(-F2+0.5d0*del)
    else
    A(m,k)=F2-0.5d0*del
    B(m)=B(m)+BCV(k)*F1
```

```
      endif
   5  continue
  10  continue

      call solver(A,B,N,1,Z)

      do 15 m=1,N
      if (BCT(m).eq.0) then
      phi(m)=BCV(m)
      dphi(m)=Z(m)
      else
      phi(m)=Z(m)
      dphi(m)=BCV(m)
      endif
  15  continue

      return
      end
```

The values of $a^{(mk)}$ in Eq. (1.30) are kept in the real array A(1:N,1:N), the sum $b^{(m1)} + b^{(m2)} + \cdots + b^{(mN)}$ on the right hand side of the equation in the real array B(1:N) and the solution $z^{(k)}$ in the real array Z(1:N). To solve for $z^{(k)}$, an LU decomposition is performed on the matrix containing the coefficients $a^{(mk)}$ to obtain a simpler system that may be easily solved by backward substitutions. This is done in the subroutine SOLVER (listed below together with supporting subprograms DAXPY, DSCAL and IDAMAX[2]) which accepts the integer N (giving the number of unknowns), the real arrays A(1:N,1:N) and B(1:N) and the integer lud to return Z(1:N). In general,

---

[2]The main part of SOLVER for decomposing the square matrix **A** and solving **AX** = **B** is respectively taken from the codes in the LINPACK subroutines DGEFA and DGESL written by Cleve Moler. The supporting subprograms DAXPY, DSCAL and IDAMAX written by Jack Dongarra are also from LINPACK. DGEFA, DGESL, DAXPY, DSCAL and IDAMAX are all in the public domain and may be downloaded from Netlib website at http://www.netlib.org. Permission for reproducing the codes here was granted by Netlib's editor-in-chief Jack Dongarra.

the integer `lud` may be given any value except 0. Neverthe-less, if we are solving two different systems of linear algebraic equations with the same square matrice $[a^{(mk)}]$, one after the other, `lud` may be given the value 0 the second time `SOLVER` is called. This is because it is not necessary to perform the LU decomposition on the same square matrix again to solve the second system after solving the first. If `lud` is given the value 0, `SOLVER` assumes that the square matrix has already been properly decomposed before and avoids the time consum-ing decomposition process. In `CELAP1`, since the square matrix has not been decomposed yet, the value of 1 is passed into `lud` when we call `SOLVER`.

The subroutine `SOLVER` and its supporting programs are listed as follows.

```
      subroutine SOLVER(A,B,N,lud,Z)

      integer lda,N,ipvt(1000),info,lud,IDAMAX,
    & j,k,kp1,l,nm1,kb

      double precision A(1000,1000),B(1000),Z(1000),
    & t,AMD(1000,1000)

      common /ludcmp/ipvt,AMD

      nm1=N-1

      do 5 i=1,N
      Z(i)=B(i)
    5 continue

      if (lud.eq.0) goto 99

      do 6 i=1,N
      do 6 j=1,N
      AMD(i,j)=A(i,j)
    6 continue
```

```fortran
      info=0

      if (nm1.lt.1) go to 70

      do 60 k=1,nm1
      kp1=k+1
      l=IDAMAX(N-k+1,AMD(k,k),1)+k-1
      ipvt(k)=l
      if (AMD(l,k).eq.0.0d0) goto 40
      if (l.eq.k) goto 10
      t=AMD(l,k)
      AMD(l,k)=AMD(k,k)
      AMD(k,k)=t
   10 continue
      t=-1.0d0/AMD(k,k)
      call DSCAL(N-k,t,AMD(k+1,k),1)
      do 30 j=kp1,N
      t=AMD(l,j)
      if (l.eq.k) go to 20
      AMD(l,j)=AMD(k,j)
      AMD(k,j)=t
   20 continue
      call DAXPY(N-k,t,AMD(k+1,k),1,AMD(k+1,j),1)
   30 continue
      goto 50
   40 continue
      info=k
   50 continue
   60 continue

   70 continue

      ipvt(N)=N

      if (AMD(N,N).eq.0.0d0) info=N
      if (info.ne.0)
```

31

```
    & pause 'Division by zero in SOLVER!'

99 continue

    if (nm1.lt.1) goto 130

    do 120 k=1,nm1
    l=ipvt(k)
    t=Z(l)
    if (l.eq.k) goto 110
    Z(l)=Z(k)
    Z(k)=t
110 continue
    call DAXPY(N-k,t,AMD(k+1,k),1,Z(k+1),1)
120 continue

130 continue

    do 140 kb=1,N
    k=N+1-kb
    Z(k) = Z(k)/AMD(k,k)
    t=-Z(k)
    call DAXPY(k-1,t,AMD(1,k),1,Z(1),1)
140 continue

    return
    end

    subroutine DAXPY(N,da,dx,incx,dy,incy)

    double precision dx(1000),dy(1000),da

    integer i,incx,incy,ix,iy,m,mp1,N

    if(N.le.0) return
    if (da .eq.  0.0d0) return
    if(incx.eq.1.and.incy.eq.1) goto 20
```

```fortran
      ix=1
      iy=1

      if(incx.lt.0) ix=(-N+1)*incx+1
      if(incy.lt.0) iy=(-N+1)*incy+1

      do 10 i=1,N
      dy(iy)=dy(iy)+da*dx(ix)
      ix=ix+incx
      iy=iy+incy
   10 continue

      return

   20 m=mod(N,4)

      if( m.eq.  0 ) go to 40

      do 30 i=1,m
      dy(i)=dy(i)+da*dx(i)
   30 continue

      if(N.lt.4) return

   40 mp1=m+1

      do 50 i=mp1,N,4
      dy(i)=dy(i)+da*dx(i)
      dy(i+1)=dy(i+1)+da*dx(i+1)
      dy(i+2)=dy(i+2)+da*dx(i+2)
      dy(i+3)=dy(i+3)+da*dx(i+3)
   50 continue

      return
      end
```

```fortran
      subroutine DSCAL(N,da,dx,incx)

      double precision da,dx(1000)

      integer i,incx,m,mp1,N,nincx

      if(N.le.0.or.incx.le.0) return
      if(incx.eq.1) goto 20
      nincx = N*incx

      do 10 i=1,nincx,incx
      dx(i)=da*dx(i)
   10 continue

      return

   20 m=mod(N,5)

      if(m.eq.0) goto 40

      do 30 i=1,m
      dx(i) = da*dx(i)
   30 continue

      if(N.lt.5) return

   40 mp1=m+1

      do 50 i=mp1,N,5
      dx(i)=da*dx(i)
      dx(i+1)=da*dx(i+1)
      dx(i+2)=da*dx(i+2)
      dx(i+3)=da*dx(i+3)
      dx(i+4)=da*dx(i+4)
   50 continue

      return
```

```fortran
      end

      function IDAMAX(N,dx,incx)

      double precision dx(1000),dmax

      integer i,incx,ix,N,IDAMAX

      IDAMAX = 0
      if(N.lt.1.or.incx.le.0 ) return
      IDAMAX = 1

      if(N.eq.1)return
      if(incx.eq.1) goto 20
      ix = 1
      dmax = dabs(dx(1))
      ix = ix + incx

      do 10 i=2,N
      if(dabs(dx(ix)).le.dmax) goto 5
      IDAMAX=i
      dmax=dabs(dx(ix))
 5    ix=ix+incx
 10   continue

      return

 20   dmax=dabs(dx(1))

      do 30 i=2,N
      if(dabs(dx(i)).le.dmax) goto 30
      IDAMAX=i
      dmax=dabs(dx(i))
 30   continue

      return
      end
```

Once the values of $\overline{\phi}^{(k)}$ and $\overline{p}^{(k)}$ are returned in the arrays phi(1:N) and dphi(1:N) by CELAP1, they can be used by the subroutine CELAP2 to compute the value of ϕ at any chosen point (ξ, η) in the interior of the solution domain. In the listing of CELAP2 below, xi and eta are the real variables which carry the values of ξ and η respectively. The computed value of $\phi(\xi, \eta)$ is returned in the real variable pint. Note that the subroutine CPF is called in CELAP2 to compute $\pi \mathcal{F}_1^{(k)}(\xi, \eta)$ and $\pi \mathcal{F}_2^{(k)}(\xi, \eta)$.

```
    subroutine CELAP2(N,xi,eta,xb,yb,nx,ny,lg,
& phi,dphi,pint)

    integer N,i

    double precision xi,eta,xb(1000),yb(1000),
& nx(1000),ny(1000),lg(1000),phi(1000),
& dphi(1000),pint,sum,pi,PF1,PF2

    pi=4d0*datan(1d0)
    sum=0d0

    do 10 i=1,N
    call CPF(xi,eta,xb(i),yb(i),nx(i),ny(i),
& lg(i),PF1,PF2)
    sum=sum+phi(i)*PF2-dphi(i)*PF1
10 continue

    pint=sum/pi

    return
    end
```

1.8 Numerical Examples

We now show how the subroutines CELAP1 and CELAP2 may be used to solve two specific examples of the interior boundary value problem described in Section 1.1.

Example 1.1

The solution domain is the square region $0 < x < 1, 0 < y < 1$. The boundary conditions are

$$\left. \begin{array}{ll} \phi = 0 & \text{on } x = 0 \\ \phi = \cos(\pi y) & \text{on } x = 1 \end{array} \right\} \text{ for } 0 < y < 1$$

$$\frac{\partial \phi}{\partial n} = 0 \text{ on } y = 0 \text{ and } y = 1 \text{ for } 0 < x < 1.$$

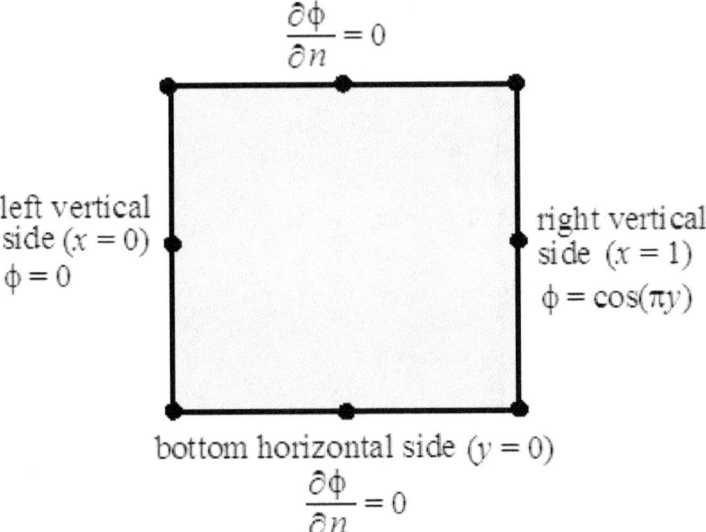

top horizontal side $(y = 1)$

$$\frac{\partial \phi}{\partial n} = 0$$

left vertical side $(x = 0)$
$\phi = 0$

right vertical side $(x = 1)$
$\phi = \cos(\pi y)$

bottom horizontal side $(y = 0)$

$$\frac{\partial \phi}{\partial n} = 0$$

Figure 1.5

The sides of the square are discretized into boundary elements of equal length. To do this, we choose N evenly spaced out points on the sides as follows. The boundary points on the sides $y = 0$ (bottom horizontal), $x = 1$ (right vertical),

$y = 1$ (top horizontal) and $x = 0$ (left vertical) are respectively given by $(x^{(m)}, y^{(m)}) = ((m-1)\ell, 0)$, $(x^{(m+N_0)}, y^{(m+N_0)}) = (1, [m-1]\ell)$, $(x^{(m+2N_0)}, y^{(m+2N_0)}) = (1 - (m-1)\ell, 1)$ and $(x^{(m+3N_0)}, y^{(m+3N_0)}) = (0, 1 - (m-1)\ell)$ for $m = 1, 2, \cdots, N_0$, where N_0 is the number of boundary elements per side (so that $N = 4N_0$) and $\ell = 1/N_0$ is the length of each element. For example, the boundary points for $N_0 = 2$ (that is, 8 boundary elements) are shown in Figure 1.5.

The input points $(x^{(1)}, y^{(1)})$, $(x^{(2)}, y^{(2)})$, $\cdots$, $(x^{(N-1)}, y^{(N-1)})$, $(x^{(N)}, y^{(N)})$ and $(x^{(N+1)}, y^{(N+1)})$, arranged in counter clockwise order on the boundary of the solution domain, are stored in the real arrays xb(1:N+1) and yb(1:N+1). (Recall that $(x^{(N+1)}, y^{(N+1)}) = (x^{(1)}, y^{(1)})$.) The values in these arrays are input data defining the geometry of the solution domain, to be generated by the user of the subroutines CELAP1 and CELAP2. As the geometry in this example is a simple one, the input data for the boundary points may be generated by writing a simple code as follows.

```
N=4*N0
dl=1d0/dfloat(N0)
do 10 i=1,N0
xb(i)=dfloat(i-1)*dl
yb(i)=0d0
xb(N0+i)=1d0
yb(N0+i)=xb(i)
xb(2*N0+i)=1d0-xb(i)
yb(2*N0+i)=1d0
xb(3*N0+i)=0d0
yb(3*N0+i)=1d0-xb(i)
10 continue
xb(N+1)=xb(1)
yb(N+1)=yb(1)
```

Note that N0 is an integer variable which gives the number of boundary elements per side and dl is a real variable giving the length of an element. The value of N0 is a given input. The boundary points in Figure 1.5 may be generated by the code above if we give N0 the value of 2.

In order to call CELAP1 and CELAP2, the midpoints of the elements (in the real arrays xm(1:N) and ym(1:N)), the lengths of the elements (in the real array lg(1:N)) and the unit normal vectors to the elements (in the real arrays nx(1:N) and ny(1:N)) are required. These can be calculated from the input data stored in the arrays xb(1:N+1) and yb(1:N+1). The general code for the calculation (which is valid for any geometry of the solution domain) is as follows.

```
      do 20 i=1,N
      xm(i)=0.5d0*(xb(i)+xb(i+1))
      ym(i)=0.5d0*(yb(i)+yb(i+1))
      lg(i)=dsqrt((xb(i+1)-xb(i))**2d0
    & +(yb(i+1)-yb(i))**2d0)
      nx(i)=(yb(i+1)-yb(i))/lg(i)
      ny(i)=(xb(i)-xb(i+1))/lg(i)
   20 continue
```

The type of boundary conditions on an element (that is, whether ϕ or $\partial\phi/\partial n$ is specified) and the corresponding specified value of either ϕ or $\partial\phi/\partial n$ are input data. The integer array BCT(1:N) is used to keep track of the types of boundary conditions on the elements. If ϕ is specified on the 5-th boundary element $C^{(5)}$ then BCT(5) is given the value 0. If BCT(5) is not 0, then we know that $\partial\phi/\partial n$ is specified on $C^{(5)}$. The values of either ϕ or $\partial\phi/\partial n$ prescribed on the boundary elements are stored in the real array BCV(1:N). For the boundary points in Figure 1.5, the input boundary values of ϕ on the two elements on the right vertical sides are given by $\cos(\pi\eta)$ with η being the y coordinates of the midpoints of the elements. For the boundary value problem here, the code for generating the input data for BCT and BCV are as follows.

```
      do 30 i=1,N
      if (i.le.N0) then
```

39

```
     BCT(i)=1
     BCV(i)=0d0
     else if ((i.gt.N0).and.(i.le.(2*N0))) then
     BCT(i)=0
     BCV(i)=dcos(pi*ym(i))
     else if ((i.gt.(2*N0)).and.(i.le.(3*N0))) then
     BCT(i)=1
     BCV(i)=0d0
     else
     BCT(i)=0
     BCV(i)=0d0
     endif
30   continue
```

We may now invoke **CELAP1** using the statement

`call CELAP1(N, xm, ym, xb, yb, nx, ny, lg, BCT, BCV, phi, dphi)`

to give us the (approximate) values of ϕ and $\partial\phi/\partial n$ on the boundary elements. The boundary values of ϕ and $\partial\phi/\partial n$ (that is, $\overline{\phi}^{(k)}$ and $\overline{p}^{(k)}$) are respectively stored in the real arrays **phi(1:N)** and **dphi(1:N)**. For example, if the variable **BCT(5)** has the value 0, we know that ϕ is specified on the 5-th boundary element and hence the variable **dphi(5)** gives us the approximate value of $\partial\phi/\partial n$ on $C^{(5)}$.

Once **CELAP1** is called, we may use **CELAP2** to calculate the value of ϕ at any interior point inside the square. For example, if we wish to calculate ϕ at $(0.50, 0.70)$, we may use the call statement

`call CELAP2(N,0.50,0.70,xb,yb,nx,ny,lg,phi,dphi,pint)`

to return us the approximate value of $\phi(0.50, 0.70)$ in the real variable **pint**.

An example of a complete program for the boundary value problem presently under consideration is given below.

```fortran
      program EX1PT1

      integer N0,BCT(1000),N,i,ians

      double precision xb(1000),yb(1000),xm(1000),
     & ym(1000),nx(1000),ny(1000),lg(1000),BCV(1000),
     & phi(1000),dphi(1000),pint,dl,xi,eta,pi

      print*,'Enter number of elements
     & per side (<250):'
      read*,N0
      N=4*N0

      pi=4d0*datan(1d0)
      dl=1d0/dfloat(N0)

      do 10 i=1,N0
      xb(i)=dfloat(i-1)*dl
      yb(i)=0d0
      xb(N0+i)=1d0
      yb(N0+i)=xb(i)
      xb(2*N0+i)=1d0-xb(i)
      yb(2*N0+i)=1d0
      xb(3*N0+i)=0d0
      yb(3*N0+i)=1d0-xb(i)
10    continue
      xb(N+1)=xb(1)
      yb(N+1)=yb(1)

      do 20 i=1,N
      xm(i)=0.5d0*(xb(i)+xb(i+1))
      ym(i)=0.5d0*(yb(i)+yb(i+1))
      lg(i)=dsqrt((xb(i+1)-xb(i))**2d0
     &   +(yb(i+1)-yb(i))**2d0)
      nx(i)=(yb(i+1)-yb(i))/lg(i)
      ny(i)=(xb(i)-xb(i+1))/lg(i)
```

41

```
20 continue

   do 30 i=1,N
   if (i.le.N0) then
   BCT(i)=1
   BCV(i)=0d0
   else if ((i.gt.N0).and.(i.le.(2*N0))) then
   BCT(i)=0
   BCV(i)=dcos(pi*ym(i))
   else if ((i.gt.(2*N0)).and.(i.le.(3*N0))) then
   BCT(i)=1
   BCV(i)=0d0
   else
   BCT(i)=0
   BCV(i)=0d0
   endif
30 continue

   call CELAP1(N,xm,ym,xb,yb,nx,ny,lg,
 &   BCT,BCV,phi,dphi)

50 print*,'Enter coordinates xi and eta of
 &   an interior point:'

   read*,xi,eta

   call CELAP2(N,xi,eta,xb,yb,nx,ny,lg,
 &   phi,dphi,pint)

   write(*,60)pint,(dexp(pi*xi)-dexp(-pi*xi))
 &   *dcos(pi*eta)/(dexp(pi)-dexp(-pi))
60 format('Numerical and exact values are:',
 &   F14.6,' and',F14.6,' respectively')

   print*,'To continue with another point
 &   enter 1:'
   read*,ians
```

```
if (ians.eq.1) goto 50

end
```

All the subprograms needed for compiling EX1PT1 into an executable program are the subroutines CELAP1, CELAP2, CPF and SOLVER (together with its supporting subprograms DAXPY, DSCAL and IDAMAX).

It is easy to check that boundary value problem here has the exact solution

$$\phi = \frac{\sinh(\pi x) \cos(\pi y)}{\sinh(\pi)}.$$

In the program EX1PT1 above, the numerical value of ϕ (as calculated by the boundary element procedure with constant elements) at an input interior point (ξ, η) is compared with the exact solution.

Table 1.1

(ξ, η)	20 elements	80 elements	Exact
$(0.10, 0.20)$	0.022605	0.022397	0.022371
$(0.10, 0.30)$	0.016454	0.016279	0.016254
$(0.10, 0.40)$	0.008681	0.008560	0.008545
$(0.50, 0.20)$	0.163153	0.161521	0.161212
$(0.50, 0.30)$	0.118290	0.117325	0.117127
$(0.50, 0.40)$	0.062107	0.061673	0.061577
$(0.90, 0.20)$	0.586250	0.590103	0.589941
$(0.90, 0.30)$	0.427451	0.428609	0.428618
$(0.90, 0.40)$	0.223159	0.225308	0.225338

The numerical values of ϕ at various interior points obtained by EX1PT1 using 20 and 80 boundary elements are compared with the exact solution in Table 1.1. There is a significant improvement in the accuracy of the numerical results when the number of boundary elements used is increased from 20 to 80.

We also examine the accuracy of the numerical value of ϕ at the interior point (a, a) as a approaches 1 from below, that is, as the point (a, a) gets closer and closer to the point $(1, 1)$ on the boundary of the square domain. The percentage errors in the numerical values of ϕ from calculations using 20 and 80 boundary elements are shown in Table 1.2 for various values of a. In each of the two sets of results, it is interesting to note that the percentage error grows as a approaches 1. For a fixed value of a near 1, the percentage error of the numerical value of ϕ calculated with 80 elements are lower than that obtained using 20 elements. It is a well known fact that the accuracy of a boundary element solution may deteriorate significantly at a point whose distance from the boundary is very small compared with the lengths of nearby boundary elements.

Table 1.2

a	0.900	0.950	0.990	0.995	0.999
20 elements	0.14%	2.83%	8.50%	9.56%	10.60%
80 elements	0.11%	0.14%	0.72%	1.40%	2.21%

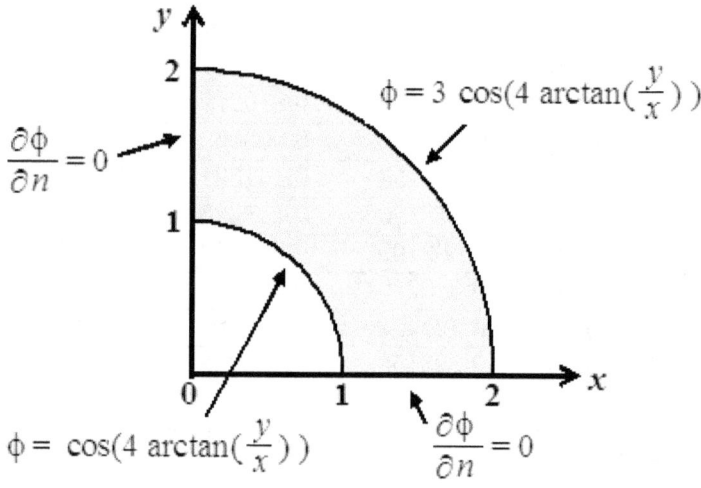

Figure 1.6

44

Example 1.2

Take the solution domain to be the region bounded between the circles $x^2 + y^2 = 1$ and $x^2 + y^2 = 4$ in the first quadrant of the Oxy plane as shown in Figure 1.6. The boundary conditions are given by

$$\frac{\partial \phi}{\partial n} = 0 \text{ on the straight side } x = 0, \ 1 < y < 2,$$

$$\frac{\partial \phi}{\partial n} = 0 \text{ on the straight side } y = 0, \ 1 < x < 2,$$

$$\phi = \cos(4\arctan(\frac{y}{x})) \text{ on the arc } x^2 + y^2 = 1, \ x > 0, \ y > 0,$$

$$\phi = 3\cos(4\arctan(\frac{y}{x})) \text{ on the arc } x^2 + y^2 = 4, \ x > 0, \ y > 0.$$

This boundary value problem may be solved numerically using the boundary element procedure with constant elements as in Example 1.1. To do this, we only have to modify the parts in the program EX1PT1 that generate input data for the arrays xb(1:N+1), yb(1:N+1), BCT(1:N) and BCV(1:N). Before we modify the program, we have to work out formulae for the boundary points $(x^{(1)}, y^{(1)})$, $(x^{(2)}, y^{(2)})$, $\cdots$, $(x^{(N-1)}, y^{(N-1)})$ and $(x^{(N)}, y^{(N)})$.

Let us discretize each of the straight sides of the boundary into N_0 elements and the arcs on $x^2 + y^2 = 1$ and $x^2 + y^2 = 4$ into $2N_0$ and $8N_0$ elements respectively, so that $N = 12N_0$. Specifically, the boundary points are given by

$$(x^{(m)}, y^{(m)}) = (1 + \frac{[m-1]}{N_0}, 0) \text{ for } m = 1, 2, \cdots, N_0,$$

$$(x^{(m+N_0)}, y^{(m+N_0)}) = (2\cos(\frac{[m-1]\pi}{16N_0}), 2\sin(\frac{[m-1]\pi}{16N_0}))$$
$$\text{for } m = 1, 2, \cdots, 8N_0,$$

$$(x^{(m+9N_0)}, y^{(m+9N_0)}) = (0, 2 - \frac{[m-1]}{N_0})$$
$$\text{for } m = 1, 2, \cdots, N_0,$$

$$(x^{(m+10N_0)}, y^{(m+10N_0)}) = (\sin(\frac{[m-1]\pi}{4N_0}), \cos(\frac{[m-1]\pi}{4N_0}))$$
$$\text{for } m = 1, 2, \cdots, 2N_0.$$

Thus, for the boundary value problem presently under consideration, the code for generating the input data for the boundary points in the real arrays xb(1:N+1) and yb(1:N+1) is as given below. Note that we are required to supply an input value for the integer N0.

```
N=12*N0
pi=4d0*datan(1d0)

do 10 i=1,8*N0
dl=pi/dfloat(16*N0)
xb(i+N0)=2d0*dcos(dfloat(i-1)*dl)
yb(i+N0)=2d0*dsin(dfloat(i-1)*dl)
if (i.le.N0) then
dl=1d0/dfloat(N0)
xb(i)=1d0+dfloat(i-1)*dl
yb(i)=0d0
xb(i+9*N0)=0d0
yb(i+9*N0)=2d0-dfloat(i-1)*dl
endif
if (i.le.(2*N0)) then
dl=pi/dfloat(4*N0)
xb(i+10*N0)=dsin(dfloat(i-1)*dl)
yb(i+10*N0)=dcos(dfloat(i-1)*dl)
endif
10 continue
xb(N+1)=xb(1)
yb(N+1)=yb(1)
```

The code for generating the input data for the integer array BCT(1:N) and the real array BCV(1:N) is as given below.

```
do 30 i=1,N
if ((i.le.N0).or.((i.gt.(9*N0))
& .and.(i.le.(10*N0)))) then
BCT(i)=1
BCV(i)=0d0
```

```
else if ((i.gt.N0).and.(i.le.(9*N0))) then
BCT(i)=0
BCV(i)=3d0*dcos(4d0*datan(ym(i)/xm(i)))
else
BCT(i)=0
BCV(i)=dcos(4d0*datan(ym(i)/xm(i)))
endif
30 continue
```

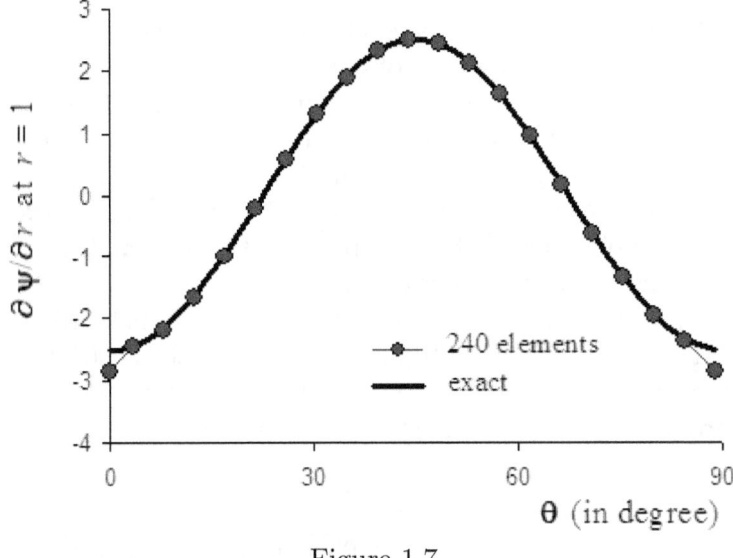

Figure 1.7

As ϕ is specified on the arc $x^2+y^2 = 1$, $x > 0$, $y > 0$, the last $2N_0$ variables in the array dphi(1:N) returned by CELAP1 give us the numerical values of $\partial\phi/\partial n$ at the midpoints of the last $2N_0$ boundary elements, that is, $-\partial\psi/\partial r$ at those midpoints if we define $\psi(r,\theta) = \phi(r\cos\theta, r\sin\theta)$, where the polar coordinates r and θ are given by by $x = r\cos\theta$ and $y = r\sin\theta$. We may print out these variables to obtain the approximate values of $\partial\psi/\partial r$ at the midpoints of the last $2N_0$ boundary elements. In Figure 1.7, the numerical $\partial\psi/\partial r$ at $r = 1$, $0 < \theta < \pi/2$, obtained using 240 elements (that is, using $N_0 = 20$) are compared graphically against the values obtained from the exact

47

solution[3] given by

$$\phi = [\frac{16}{85}([x^2+y^2]^2 - \frac{1}{[x^2+y^2]^2}) - \frac{16}{255}(\frac{[x^2+y^2]^2}{16}$$
$$- \frac{16}{[x^2+y^2]^2})] \cos(4\arctan(\frac{y}{x})).$$

The numerical values show a good agreement with the exact ones except at points that are extremely close to the corner points $(0,1)$ and $(1,0)$, that is, except at near $\theta = 0$ and $\theta = \pi/2$.

The numerical values of ϕ at selected points in the interior of the solution domain, obtained using 240 elements, are compared with the exact solution in Table 1.3. There is a good agreement between the two sets of results. The interior points in the last two rows of Table 1.3 are close to the corner point $(1,0)$. Note that the errors of the numerical values at these two points are higher compared with those at the other points. When we repeat the same calculation using 480 elements ($N_0 = 40$), the numerical values of ϕ are 0.826108 and 0.974111 at $(1.099998, 0.001920)$ and $(1.010000, 0.000176)$ respectively, that is, we observe a significant improvement in the accuracy of the numerical values at the two points.

Table 1.3

(ξ, η)	240 elements	Exact
$(1.082532, 0.625000)$	-0.392546	-0.392045
$(0.875000, 1.515544)$	-0.908254	-0.907816
$(1.060660, 1.060660)$	-1.094489	-1.094211
$(1.099998, 0.001920)$	0.824548	0.826958
$(1.010000, 0.000176)$	0.960174	0.975656

1.9   Summary and Discussion

A boundary element solution for the interior boundary value problem defined by Eqs. (1.1)-(1.2) is given by Eq. (1.32) to-

---

[3]Refer to page 202 of the book *Partial Differential Equations in Mechanics 1* by APS Selvadurai (Springer-Verlag, 2000).

48

gether with Eqs. (1.28), (1.30) and (1.31). The solution is constructed from the boundary integral solution in Eq. (1.23). Constant elements are used, that is, the boundary (of the solution domain) is discretized into straight line elements and the solution ϕ and its normal derivative $\partial\phi/\partial n$ on the boundary are approximated as constants over a boundary element.

As no discretization of the entire solution domain is required, the boundary element solution may be easily implemented on the computer for problems involving complicated geometries and general boundary conditions. The boundary may be easily discretized into line elements by merely placing on it well spaced out points. We have discussed in detail how the numerical procedure can be coded in FORTRAN 77. In spite of the specific programming language used, our discussion may still be useful to readers who are interested in developing the method using other software tools (such as C++ and MATLAB), as FORTRAN 77 codes are relatively easy to decipher.

The term "direct boundary element method" is often used to describe the boundary element procedure given in this chapter. This is because the unknowns in the formulation given by Eq. (1.30) can be directly interpreted as values of ϕ or $\partial\phi/\partial n$ on the boundary. An alternative boundary element method may be obtained from the simpler boundary integral solution

$$\phi(x,y) = \int\limits_{C} A(\xi,\eta)\ln([x-\xi]^2 + [y-\eta]^2)ds(\xi,\eta),$$

where $A(\xi,\eta)$ is a (boundary) weight function yet to be determined. To determine $A(\xi,\eta)$ approximately, we discretize C into boundary elements $C^{(1)}, C^{(2)}, \cdots, C^{(N-1)}$ and $C^{(N)}$ as before, and approximate $A(\xi,\eta)$ as a constant $A^{(m)}$ over $C^{(m)}$, in order to obtain the approximation

$$\phi(x,y) \simeq \sum_{m=1}^{N} A^{(m)} \int\limits_{C^{(m)}} \ln([x-\xi]^2 + [y-\eta]^2)ds(\xi,\eta).$$

The constants $A^{(m)}$ are to be determined by using the given boundary conditions. We shall not go into further details here other than pointing out that such as an approach gives rise to a so called indirect boundary element method as the unknowns $A^{(m)}$ are not related to ϕ or $\partial\phi/\partial n$ on the boundary in a simple and direct manner.

1.10 Exercises

1. If ϕ satisfies the two-dimensional Laplace's equation in the region R bounded by a simple closed curve C, use the divergence theorem to show that

$$\int_C \frac{\partial}{\partial n}[\phi(x,y)]ds(x,y) = 0.$$

(Note. This implies that if we prescribe $\partial\phi/\partial n$ at all points on C in our boundary value problem we have to be careful to ensure the above equation is satisfied. Otherwise, the boundary value problem does not have a solution.)

2. If ϕ satisfies the two-dimensional Laplace's equation in the region R bounded by the curve C, use the divergence theorem to derive the relation

$$\iint_R |\underline{\nabla}\phi(x,y)|^2 dxdy = \int_C \phi(x,y)\frac{\partial}{\partial n}[\phi(x,y)]ds(x,y).$$

Hence show that: (a) if $\phi = 0$ at all points on C then $\phi = 0$ at all points in R, that is, show that if the boundary conditions are given by $\phi = 0$ on C then the solution of our boundary value problem is uniquely given by $\phi = 0$ for $(x,y) \in R$, and (b) if $\partial\phi/\partial n = 0$ at all points on C then ϕ can be any arbitrary constant function in R, that is, if the boundary conditions are given by $\partial\phi/\partial n = 0$ on C, then our boundary value problem has infinitely many solutions given by $\phi = c$ for $(x,y) \in R$, where c is an arbitrary constant.

3. Use the result in Exercise 2(a) above to show that if the boundary conditions are given by $\phi = f(x,y)$ at all points on the simple closed curve C then the boundary value problem governed by the two-dimensional Laplace's equation in the region R has a unique solution. [Hint. Show that if ϕ_1 and ϕ_2 are any two solutions satisfying the Laplace's equation and the boundary conditions under consideration then $\phi_1 = \phi_2$ at all points in R.] (Notes. (1) In general, for the interior boundary value problem defined by Eqs. (1.1)-(1.2) to have a unique solution, ϕ must be specified at *at least one* point on C. (2) For the case in which $\partial\phi/\partial n$ is specified at all points on C, ϕ is only determined to within an arbitrary constant. In such a case, the boundary element procedure in this chapter may still work to give us one of the infinitely many solutions.)

4. Eq. (1.8) is not the only solution of the two-dimensional Laplace's equation that is not well defined at the single point (ξ, η). By differentiating Eq. (1.8) partially with respect to x and/or y as many times as we like, we may generate other solutions that are not well defined at (ξ, η). An example of these other solutions is

$$\phi(x,y) = \frac{(x - \xi)}{2\pi[(x - \xi)^2 + (y - \eta)^2]}.$$

If we denote this solution by $\Phi(x, y; \xi, \eta)$ (like what we had done before for the solution in Eq. (1.8)), investigate whether we can still derive the boundary integral solution as given by Eq. (1.19) from the reciprocal relation in Eq. (1.10) or not.

5. Explain why the parameter $\lambda(\xi, \eta)$ in Eq. (1.23) can be calculated using

$$\lambda(\xi, \eta) = \int\limits_C \frac{\partial}{\partial n}(\Phi(x, y; \xi, \eta))ds(x, y).$$

Taking C to be the boundary of the triangular region $y < -x + 1$, $x > 0$, $y > 0$, evaluate the line integral above to check that: (a) $\lambda(2,1) = 0$, (b) $\lambda(1,0) = 1/8$, (c) $\lambda(0,0) = 1/4$, (d) $\lambda(1/2, 1/2) = 1/2$, and (e) $\lambda(1/2, 1/4) = 1$.

6. The boundary element solution given in this chapter provides us with an approximate but explicit formula for calculating ϕ at any interior point (ξ, η) in the solution domain. We may also be interested in computing the vector quantity $\nabla\phi$. Can an approximate explicit formula be obtained for $\nabla\phi$ at (ξ, η)? How can we obtain one?

7. Modify the program EX1PT1 in Section 1.7 to solve numerically the Laplace's equation given by Eq. (1.1) in the region $x^2 + y^2 < 1$, $x > 0$, $y > 0$, subject to the boundary conditions

$$\phi = \text{0 on the horizontal side } y = 0 \text{ for } 0 < x < 1,$$
$$\phi = \text{1 on the vertical side } x = 0 \text{ for } 0 < y < 1,$$
$$\frac{\partial \phi}{\partial n} = \text{0 on the quarter circle}$$
$$x^2 + y^2 = 1, \ x > 0, \ y > 0.$$

Discretize each of the three parts of the boundary (the two sides and the quarter circle) into N_0 boundary elements so that $N = 3N_0$, that is, so that we have a total of $3N_0$ boundary elements. Using various values of N_0, run the modified program to calculate the numerical solution at various selected interior points and compare the results obtained with the exact solution $\phi(x,y) = (2/\pi) \arctan(y/x)$. According to the exact solution, ϕ is not well defined at $(0,0)$. This is not surprising, as $(0,0)$ is the point where the value of ϕ suddenly jumps from 0 on the horizontal side to 1 on the vertical side. The normal derivative $\partial \phi / \partial n$ is given by $-\partial \phi / \partial y$ on the horizontal side. From the exact solution, we know that $\partial \phi / \partial n \to -\infty$ on $y = 0$ as $x \to 0^+$. This singular behavior of the solution may be a problem for the boundary element procedure. Investigate the numerical solution at points near

$(0,0)$. The subroutine CELAP1 returns the numerical values of $\partial\phi/\partial n$ on the horizontal side in the first N_0 variables of the real array dphi(1:N). Compare the values in these variables with the exact values of $\partial\phi/\partial n$ on the horizontal side. Repeat the same exercise with the vertical side.

8. Modify the program EX1PT1 in Section 1.7 to solve numerically the Laplace's equation given by Eq. (1.1) in the region $0 < x < 1$, $0 < y < 1$, subject to the boundary conditions

$$\frac{\partial\phi}{\partial n} = 5\cos(\pi x) \text{ on the top horizontal}$$
$$\text{side } y = 1 \text{ for } 0 < x < 1,$$
$$\frac{\partial\phi}{\partial n} = 0 \text{ on the other three remaining sides,}$$
$$\phi = 1 \text{ at } (x,y) = (\frac{1}{3}, \frac{1}{2}).$$

By discretizing each side of the square into N_0 equal length elements, run the modified program with various values of N_0 to compute ϕ at some selected interior points. (Notes. (1) With the boundary conditions alone, the boundary value problem does not have a unique solution. Thus, the subroutine CELAP2 by itself may not return us the desired numerical value of ϕ in the real variable pint. Use the additional condition as given by $\phi(1/3, 1/2) = 1$ to find the desired numerical solution. (2) The exact solution of this problem is given by $\phi(x,y) = 5\cos(\pi x)\cosh(\pi y)/(\pi \sinh(\pi)) + 0.827\,103\,138\,436$. Compare your numerical values with the exact solution.)

Chapter 2

Discontinuous Linear Elements

2.1 Introduction

In Chapter 1, we have studied a simple boundary element method for solving numerically the Laplace's equation

$$\frac{\partial^2 \phi}{\partial x^2} + \frac{\partial^2 \phi}{\partial y^2} = 0 \qquad (2.1)$$

in the two-dimensional region R bounded by a closed curve C, given that either ϕ or its (outward) normal derivative $\partial \phi / \partial n$ is suitably specified at each and every point on C.

The method is derived from the boundary integral equation

$$\lambda(\xi, \eta)\phi(\xi, \eta) = \int_C [\phi(x,y)\frac{\partial}{\partial n}(\Phi(x,y;\xi,\eta))$$

$$-\Phi(x,y;\xi,\eta)\frac{\partial}{\partial n}(\phi(x,y))]ds(x,y) \quad (2.2)$$

by using constant elements, that is, through the approximation

$$C \simeq C^{(1)} \cup C^{(2)} \cup \cdots \cup C^{(N-1)} \cup C^{(N)},$$

$$\phi \simeq \overline{\phi}^{(k)} \text{ and } \frac{\partial \phi}{\partial n} \simeq \overline{p}^{(k)} \text{ for } (x,y) \in C^{(k)} \ (k = 1, 2, \cdots, N),$$

$$(2.3)$$

where $C^{(1)}$, $C^{(2)}$, $\cdots$, $C^{(N-1)}$ and $C^{(N)}$ are straight line segments and $\overline{\phi}^{(k)}$ and $\overline{p}^{(k)}$ are respectively the values of ϕ and $\partial\phi/\partial n$ at the midpoint of $C^{(k)}$.

To obtain more accurate results, the boundary element calculation can always be refined by employing more elements. Nevertheless, there is a practical limitation to the number of elements that may be used. If the number of elements is too large, the setting up and the solution of the system of linear algebraic equations may become computationally time and memory consuming.

If a boundary element technique that can give better results with fewer elements is needed, we may attempt to devise higher order elements by improving the approximation in Eq. (2.3). We may allow the approximation of ϕ and $\partial\phi/\partial n$ to vary across a boundary element. Furthermore, if some part of the boundary is a curve, it may be better approximated using small curved segments rather than straight line ones. All this may, however, give rise to a very complicated formulation which may not be as easy to implement as the constant elements. We shall not embark on a general studies of higher order elements here. The specific aim of this chapter is merely to introduce the idea of discontinuous linear elements and show how these elements may be employed to solve the interior boundary value problem governed by Eq. (2.1).

In devising linear elements, the boundary of the solution domain is discretized into straight line segments as in the case of constant elements in Chapter 1. The function ϕ and its normal derivative $\partial\phi/\partial n$ are approximated by using functions that vary linearly across the boundary elements. Continuous linear elements are obtained if we constrain the approximate linear variations of ϕ and $\partial\phi/\partial n$ to be continuous from one element to another (at the common endpoints). Continuous linear elements do not always perform as well as one may expect. One reason for this is that $\partial\phi/\partial n$ is not continuous at sharp corners (if any) on the boundary of the solution domain. Furthermore, in some problems, ϕ may also be discontinuous at certain points on the boundary. On the whole, discontinuous linear elements which do not require the approximate linear variations of ϕ and $\partial\phi/\partial n$ to be continuous at the endpoints of the boundary elements are known to give more accurate numerical solutions than constant or continuous linear elements.

2.2 Boundary Element Solution with Discontinuous Linear Elements

As in Chapter 1, we discretize the boundary C of the solution domain into N straight line segments $C^{(1)}$, $C^{(2)}$, $\cdots$, $C^{(N-1)}$ and $C^{(N)}$ by placing on it N well spaced out points $(x^{(1)}, y^{(1)})$, $(x^{(2)}, y^{(2)})$, $\cdots$, $(x^{(N-1)}, y^{(N-1)})$ and $(x^{(N)}, y^{(N)})$ in the counter clockwise order. The endpoints of the boundary element $C^{(k)}$ are $(x^{(k)}, y^{(k)})$ and $(x^{(k+1)}, y^{(k+1)})$. (Note that we define $(x^{(N+1)}, y^{(N+1)}) = (x^{(1)}, y^{(1)})$.)

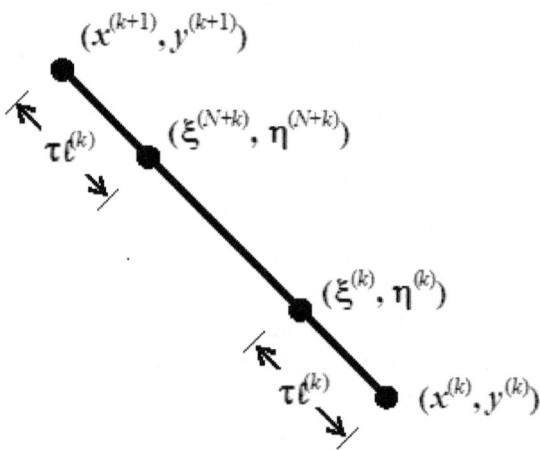

Figure 2.1

In Eq. (2.3), across the entire element $C^{(k)}$, we approximate ϕ and $\partial\phi/\partial n$ by their values at the midpoint of $C^{(k)}$. To construct a linear approximation, we need to use the values of ϕ and $\partial\phi/\partial n$ at two distinct chosen points on the element. On $C^{(k)}$, we choose the two points $(\xi^{(k)}, \eta^{(k)})$ and $(\xi^{(N+k)}, \eta^{(N+k)})$ to be at a distance of $\tau\ell^{(k)}$ from the endpoints $(x^{(k)}, y^{(k)})$ and $(x^{(k+1)}, y^{(k+1)})$ respectively, where τ is a positive number such that $0 < \tau < 1/2$ and $\ell^{(k)}$ is the length of $C^{(k)}$. Refer to Figure 2.1.

Let the values of ϕ at $(\xi^{(k)}, \eta^{(k)})$ and $(\xi^{(N+k)}, \eta^{(N+k)})$ be denoted by $\widehat{\phi}^{(k)}$ and $\widehat{\phi}^{(N+k)}$ respectively. To derive an approximate linear variation for ϕ across $C^{(k)}$ in terms of $\widehat{\phi}^{(k)}$ and $\widehat{\phi}^{(N+k)}$, we define

$$s(x,y) = \sqrt{(x - x^{(k)})^2 + (y - y^{(k)})^2} \ \text{ for } (x,y) \in C^{(k)}, \quad (2.4)$$

that is, $s(x,y)$ gives the distance between the point (x,y) on $C^{(k)}$ and the endpoint $(x^{(k)}, y^{(k)})$ of the element. Note that $s(x,y) = \tau\ell^{(k)}$ and $s(x,y) = (1-\tau)\ell^{(k)}$ at $(x,y) = (\xi^{(k)}, \eta^{(k)})$ and $(x,y) = (\xi^{(N+k)}, \eta^{(N+k)})$ respectively. The required linear approximation for ϕ in terms of $\widehat{\phi}^{(k)}$ and $\widehat{\phi}^{(N+k)}$ is then given by

$$\phi(x,y)$$
$$\simeq \frac{[s(x,y) - (1-\tau)\ell^{(k)}]\widehat{\phi}^{(k)} - [s(x,y) - \tau\ell^{(k)}]\widehat{\phi}^{(N+k)}}{(2\tau - 1)\ell^{(k)}}$$
$$\text{for } (x,y) \in C^{(k)}. \quad (2.5)$$

Similarly, denoting the values of $\partial\phi/\partial n$ at $(\xi^{(k)}, \eta^{(k)})$ and $(\xi^{(N+k)}, \eta^{(N+k)})$ by $\widehat{p}^{(k)}$ and $\widehat{p}^{(N+k)}$ respectively, we make the approximation

$$\frac{\partial}{\partial n}[\phi(x,y)]$$
$$\simeq \frac{[s(x,y) - (1-\tau)\ell^{(k)}]\widehat{p}^{(k)} - [s(x,y) - \tau\ell^{(k)}]\widehat{p}^{(N+k)}}{(2\tau - 1)\ell^{(k)}}$$
$$\text{for } (x,y) \in C^{(k)}. \quad (2.6)$$

For $0 < \tau < 1/2$, the approximation in Eqs. (2.5) and (2.6) defines the discontinuous linear elements. We assume that

C is discretized in such a way that either ϕ or $\partial\phi/\partial n$ (not both) is specified over a boundary element. If ϕ is specified over $C^{(k)}$ then $\widehat{p}^{(k)}$ and $\widehat{p}^{(N+k)}$ are unknown constants. On the other hand, if $\partial\phi/\partial n$ is given over $C^{(k)}$, $\widehat{\phi}^{(k)}$ and $\widehat{\phi}^{(N+k)}$ are unknowns. Thus, there are $2N$ unknowns to be determined in Eqs. (2.5) and (2.6).

Substitution of Eqs. (2.5) and (2.6) into Eq. (2.2) yields

$$
\begin{aligned}
\lambda(\xi,\eta)&\phi(\xi,\eta) \\
\simeq \sum_{k=1}^{N} & \frac{1}{(2\tau-1)\ell^{(k)}} \\
\times \Big\{ & \widehat{\phi}^{(k)}[-(1-\tau)\ell^{(k)}\mathcal{F}_2^{(k)}(\xi,\eta) + \mathcal{F}_4^{(k)}(\xi,\eta)] \\
& +\widehat{\phi}^{(N+k)}[\tau\ell^{(k)}\mathcal{F}_2^{(k)}(\xi,\eta) - \mathcal{F}_4^{(k)}(\xi,\eta)] \\
& -\widehat{p}^{(k)}[-(1-\tau)\ell^{(k)}\mathcal{F}_1^{(k)}(\xi,\eta) + \mathcal{F}_3^{(k)}(\xi,\eta)] \\
& -\widehat{p}^{(N+k)}[\tau\ell^{(k)}\mathcal{F}_1^{(k)}(\xi,\eta) - \mathcal{F}_3^{(k)}(\xi,\eta)] \Big\},
\end{aligned}
\tag{2.7}
$$

where

$$
\begin{aligned}
\mathcal{F}_1^{(k)}(\xi,\eta) &= \int_{C^{(k)}} \Phi(x,y;\xi,\eta)ds(x,y), \\
\mathcal{F}_2^{(k)}(\xi,\eta) &= \int_{C^{(k)}} \frac{\partial}{\partial n}[\Phi(x,y;\xi,\eta)]ds(x,y), \\
\mathcal{F}_3^{(k)}(\xi,\eta) &= \int_{C^{(k)}} s(x,y)\Phi(x,y;\xi,\eta)ds(x,y), \\
\mathcal{F}_4^{(k)}(\xi,\eta) &= \int_{C^{(k)}} s(x,y)\frac{\partial}{\partial n}[\Phi(x,y;\xi,\eta)]ds(x,y). \quad (2.8)
\end{aligned}
$$

To set up a system of $2N$ linear algebraic equations to determine the $2N$ unknowns on the right hand side of Eq. (2.7), we let (ξ,η) be given in turn by $(\xi^{(m)},\eta^{(m)})$ for $m=1,2,\cdots,$

$2N$ to obtain

$$
\begin{aligned}
\frac{1}{2}\widehat{\phi}^{(m)} \simeq \sum_{k=1}^{N} & \frac{1}{(2\tau-1)\ell^{(k)}} \\
\times \Big\{ & \widehat{\phi}^{(k)}[-(1-\tau)\ell^{(k)}\mathcal{F}_2^{(k)}(\xi^{(m)},\eta^{(m)}) \\
& +\mathcal{F}_4^{(k)}(\xi^{(m)},\eta^{(m)})] \\
& +\widehat{\phi}^{(N+k)}[\tau\ell^{(k)}\mathcal{F}_2^{(k)}(\xi^{(m)},\eta^{(m)})-\mathcal{F}_4^{(k)}(\xi^{(m)},\eta^{(m)})] \\
& -\widehat{p}^{(k)}[-(1-\tau)\ell^{(k)}\mathcal{F}_1^{(k)}(\xi^{(m)},\eta^{(m)}) \\
& +\mathcal{F}_3^{(k)}(\xi^{(m)},\eta^{(m)})] \\
& -\widehat{p}^{(N+k)}[\tau\ell^{(k)}\mathcal{F}_1^{(k)}(\xi^{(m)},\eta^{(m)})-\mathcal{F}_3^{(k)}(\xi^{(m)},\eta^{(m)})] \Big\} \\
& \text{for } m=1,2,\cdots,2N. \qquad (2.9)
\end{aligned}
$$

Note that $(\xi^{(k)},\eta^{(k)})$ and $(\xi^{(N+k)},\eta^{(N+k)})$ are two points lying on $C^{(k)}$, somewhere in between the endpoints of the element. Thus, we may take $\lambda(\xi^{(m)},\eta^{(m)})=1/2$ for $m=1,2,\cdots,2N$.

The system of $2N$ linear algebraic equations containing $2N$ unknowns given by Eq. (2.9) may be rewritten as

$$
\sum_{k=1}^{N}(a^{(mk)}z^{(k)}+a^{(m[N+k])}z^{(N+k)})
$$
$$
= \sum_{k=1}^{N}b^{(mk)} \text{ for } m=1,2,\cdots,2N, \qquad (2.10)
$$

where

$$
a^{(mk)} = \begin{cases}
(2\tau-1)^{-1}[(1-\tau)\mathcal{F}_1^{(k)}(\xi^{(m)},\eta^{(m)}) \\
\quad -[\ell^{(k)}]^{-1}\mathcal{F}_3^{(k)}(\xi^{(m)},\eta^{(m)})] \\
\qquad \text{if } \phi \text{ is given on } C^{(k)}, \\
(2\tau-1)^{-1}[-(1-\tau)\mathcal{F}_2^{(k)}(\xi^{(m)},\eta^{(m)}) \\
\quad +[\ell^{(k)}]^{-1}\mathcal{F}_4^{(k)}(\xi^{(m)},\eta^{(m)})]-\frac{1}{2}\delta^{(mk)} \\
\qquad \text{if } \partial\phi/\partial n \text{ is given on } C^{(k)},
\end{cases}
$$

$$a^{(m[N+k])} = \begin{cases} (2\tau-1)^{-1}[-\tau\mathcal{F}_1^{(k)}(\xi^{(m)},\eta^{(m)}) \\ +[\ell^{(k)}]^{-1}\mathcal{F}_3^{(k)}(\xi^{(m)},\eta^{(m)})] \\ \qquad \text{if } \phi \text{ is given on } C^{(k)}, \\ (2\tau-1)^{-1}[\tau\mathcal{F}_2^{(k)}(\xi^{(m)},\eta^{(m)}) \\ -[\ell^{(k)}]^{-1}\mathcal{F}_4^{(k)}(\xi^{(m)},\eta^{(m)})] - \frac{1}{2}\delta^{([m-N]k)} \\ \qquad \text{if } \partial\phi/\partial n \text{ is given on } C^{(k)}, \end{cases}$$

$$b^{(mk)} = \begin{cases} -\widehat{\phi}^{(k)}\{(2\tau-1)^{-1} \\ \times[-(1-\tau)\mathcal{F}_2^{(k)}(\xi^{(m)},\eta^{(m)}) \\ +[\ell^{(k)}]^{-1}\mathcal{F}_4^{(k)}(\xi^{(m)},\eta^{(m)})] - \frac{1}{2}\delta^{(mk)}\} \\ -\widehat{\phi}^{(N+k)}\{(2\tau-1)^{-1}[\tau\mathcal{F}_2^{(k)}(\xi^{(m)},\eta^{(m)}) \\ -[\ell^{(k)}]^{-1}\mathcal{F}_4^{(k)}(\xi^{(m)},\eta^{(m)})] - \frac{1}{2}\delta^{([m-N]k)}\} \\ \qquad \text{if } \phi \text{ is given on } C^{(k)}, \\ -\widehat{p}^{(k)}(2\tau-1)^{-1}[(1-\tau)\mathcal{F}_1^{(k)}(\xi^{(m)},\eta^{(m)}) \\ -[\ell^{(k)}]^{-1}\mathcal{F}_3^{(k)}(\xi^{(m)},\eta^{(m)})] \\ -\widehat{p}^{(N+k)}(2\tau-1)^{-1}[-\tau\mathcal{F}_1^{(k)}(\xi^{(m)},\eta^{(m)}) \\ +[\ell^{(k)}]^{-1}\mathcal{F}_3^{(k)}(\xi^{(m)},\eta^{(m)})] \\ \qquad \text{if } \partial\phi/\partial n \text{ is given on } C^{(k)}, \end{cases}$$

$$\delta^{(mk)} = \begin{cases} 0 & \text{if } m \neq k, \\ 1 & \text{if } m = k, \end{cases}$$

$$z^{(k)} = \begin{cases} \widehat{p}^{(k)} & \text{if } \phi \text{ is given on } C^{(k)}, \\ \widehat{\phi}^{(k)} & \text{if } \partial\phi/\partial n \text{ is given on } C^{(k)}, \end{cases}$$

$$z^{(N+k)} = \begin{cases} \widehat{p}^{(N+k)} & \text{if } \phi \text{ is given on } C^{(k)}, \\ \widehat{\phi}^{(N+k)} & \text{if } \partial\phi/\partial n \text{ is given on } C^{(k)}. \end{cases} \tag{2.11}$$

Exact formulae for calculating the integrals which define $\mathcal{F}_1^{(k)}(\xi,\eta)$, $\mathcal{F}_2^{(k)}(\xi,\eta)$, $\mathcal{F}_3^{(k)}(\xi,\eta)$ and $\mathcal{F}_4^{(k)}(\xi,\eta)$ are given below by Eqs. (2.12), (2.13), (2.14), (2.15), (2.19) and (2.20) together with Eq. (2.16).

2.3   Formulae for Integrals of Discontinuous Linear Elements

As derived in Chapter 1, the exact formulae for $\mathcal{F}_1^{(k)}(\xi,\eta)$ and $\mathcal{F}_2^{(k)}(\xi,\eta)$ are given by

$$
\begin{aligned}
&\mathcal{F}_1^{(k)}(\xi,\eta) \\
&= \frac{\ell^{(k)}}{2\pi}\Big\{\ln(\ell^{(k)}) + \Big(1 + \frac{B^{(k)}(\xi,\eta)}{2A^{(k)}}\Big)\ln\Big|1 + \frac{B^{(k)}(\xi,\eta)}{2A^{(k)}}\Big| \\
&\quad - \frac{B^{(k)}(\xi,\eta)}{2A^{(k)}}\ln\Big|\frac{B^{(k)}(\xi,\eta)}{2A^{(k)}}\Big| - 1\Big\} \\
&\qquad \text{if } 4A^{(k)}E^{(k)}(\xi,\eta) - [B^{(k)}(\xi,\eta)]^2 = 0, \qquad (2.12)
\end{aligned}
$$

$$
\mathcal{F}_2^{(k)}(\xi,\eta) = 0 \;\text{ if }\; 4A^{(k)}E^{(k)}(\xi,\eta) - [B^{(k)}(\xi,\eta)]^2 = 0, \quad (2.13)
$$

$$
\begin{aligned}
&\mathcal{F}_1^{(k)}(\xi,\eta) \\
&= \frac{\ell^{(k)}}{4\pi}\Big\{2[\ln(\ell^{(k)}) - 1] - \frac{B^{(k)}(\xi,\eta)}{2A^{(k)}}\ln\Big|\frac{E^{(k)}(\xi,\eta)}{A^{(k)}}\Big| \\
&\quad +\Big(1 + \frac{B^{(k)}(\xi,\eta)}{2A^{(k)}}\Big)\ln\Big|1 + \frac{B^{(k)}(\xi,\eta)}{A^{(k)}} + \frac{E^{(k)}(\xi,\eta)}{A^{(k)}}\Big| \\
&\quad + \frac{\sqrt{4A^{(k)}E^{(k)}(\xi,\eta) - [B^{(k)}(\xi,\eta)]^2}}{A^{(k)}} \\
&\quad \times\Big[\arctan\Big(\frac{2A^{(k)} + B^{(k)}(\xi,\eta)}{\sqrt{4A^{(k)}E^{(k)}(\xi,\eta) - [B^{(k)}(\xi,\eta)]^2}}\Big) \\
&\quad - \arctan\Big(\frac{B^{(k)}(\xi,\eta)}{\sqrt{4A^{(k)}E^{(k)}(\xi,\eta) - [B^{(k)}(\xi,\eta)]^2}}\Big)\Big]\Big\} \\
&\qquad \text{if } 4A^{(k)}E^{(k)}(\xi,\eta) - [B^{(k)}(\xi,\eta)]^2 > 0, \quad (2.14)
\end{aligned}
$$

$$\mathcal{F}_2^{(k)}(\xi, \eta)$$

$$= \frac{\ell^{(k)}[n_x^{(k)}(x^{(k)} - \xi) + n_y^{(k)}(y^{(k)} - \eta)]}{\pi \sqrt{4A^{(k)} E^{(k)}(\xi, \eta) - [B^{(k)}(\xi, \eta)]^2}}$$

$$\times [\arctan(\frac{2A^{(k)} + B^{(k)}(\xi, \eta)}{\sqrt{4A^{(k)} E^{(k)}(\xi, \eta) - [B^{(k)}(\xi, \eta)]^2}})$$

$$- \arctan(\frac{B^{(k)}(\xi, \eta)}{\sqrt{4A^{(k)} E^{(k)}(\xi, \eta) - [B^{(k)}(\xi, \eta)]^2}})]$$

$$\text{if } 4A^{(k)} E^{(k)}(\xi, \eta) - [B^{(k)}(\xi, \eta)]^2 > 0, \qquad (2.15)$$

together with

$$A^{(k)} = [\ell^{(k)}]^2,$$
$$B^{(k)}(\xi, \eta) = [-n_y^{(k)}(x^{(k)} - \xi) + (y^{(k)} - \eta)n_x^{(k)}](2\ell^{(k)}),$$
$$E^{(k)}(\xi, \eta) = (x^{(k)} - \xi)^2 + (y^{(k)} - \eta)^2, \qquad (2.16)$$

where $[n_x^{(k)}, n_y^{(k)}] = [y^{(k+1)} - y^{(k)}, x^{(k)} - x^{(k+1)}]/\ell^{(k)}$ is the unit normal vector to $C^{(k)}$ pointing away from the solution domain.

To obtain formulae for $\mathcal{F}_3^{(k)}(\xi, \eta)$ and $\mathcal{F}_4^{(k)}(\xi, \eta)$, we parameterize points (x, y) on the boundary element $C^{(k)}$ using

$$\left. \begin{array}{l} x = x^{(k)} - t\ell^{(k)}n_y^{(k)} \\ y = y^{(k)} + t\ell^{(k)}n_x^{(k)} \end{array} \right\} \text{ from } t = 0 \text{ to } t = 1. \qquad (2.17)$$

With Eqs. (2.8) and (2.17), we find that

$$\mathcal{F}_3^{(k)}(\xi, \eta)$$

$$= \frac{[\ell^{(k)}]^2}{4\pi} \int_0^1 t \ln[A^{(k)}t^2 + B^{(k)}(\xi, \eta)t + E^{(k)}(\xi, \eta)]dt,$$

$$\mathcal{F}_4^{(k)}(\xi, \eta)$$

$$= \frac{[\ell^{(k)}]^2}{2\pi} \int_0^1 \frac{t[n_x^{(k)}(x^{(k)} - \xi) + n_y^{(k)}(y^{(k)} - \eta)]}{A^{(k)}t^2 + B^{(k)}(\xi, \eta)t + E^{(k)}(\xi, \eta)}dt. \qquad (2.18)$$

We may rewrite Eq. (2.18) as

$$
\mathcal{F}_3^{(k)}(\xi,\eta)
$$

$$
= -\frac{B^{(k)}(\xi,\eta)[\ell^{(k)}]^2}{8\pi A^{(k)}} \int_0^1 \ln[A^{(k)}t^2 + B^{(k)}(\xi,\eta)t + E^{(k)}(\xi,\eta)]dt
$$

$$
+\frac{[\ell^{(k)}]^2}{8\pi A^{(k)}} \int_0^1 [2A^{(k)}t + B^{(k)}(\xi,\eta)]
$$

$$
\times \ln[A^{(k)}t^2 + B^{(k)}(\xi,\eta)t + E^{(k)}(\xi,\eta)]dt,
$$

$$
\mathcal{F}_4^{(k)}(\xi,\eta)
$$

$$
= -\frac{B^{(k)}(\xi,\eta)[\ell^{(k)}]^2}{4\pi A^{(k)}} \int_0^1 \frac{[n_x^{(k)}(x^{(k)}-\xi)+n_y^{(k)}(y^{(k)}-\eta)]}{A^{(k)}t^2 + B^{(k)}(\xi,\eta)t + E^{(k)}(\xi,\eta)}dt
$$

$$
+\frac{[\ell^{(k)}]^2}{4\pi A^{(k)}} \int_0^1 \frac{[2A^{(k)}t + B^{(k)}(\xi,\eta)]}{A^{(k)}t^2 + B^{(k)}(\xi,\eta)t + E^{(k)}(\xi,\eta)}
$$

$$
\times [n_x^{(k)}(x^{(k)}-\xi)+n_y^{(k)}(y^{(k)}-\eta)]dt
$$

to obtain

$$
\begin{aligned}
\mathcal{F}_3^{(k)}(\xi,\eta) &= -\frac{B^{(k)}(\xi,\eta)\ell^{(k)}}{2A^{(k)}}\mathcal{F}_1^{(k)}(\xi,\eta) \\
&\quad +\frac{[\ell^{(k)}]^2}{8\pi A^{(k)}}\{(A^{(k)} + B^{(k)}(\xi,\eta) + E^{(k)}(\xi,\eta)) \\
&\quad \times (\ln|A^{(k)} + B^{(k)}(\xi,\eta) + E^{(k)}(\xi,\eta)| - 1) \\
&\quad - E^{(k)}(\xi,\eta)(\ln|E^{(k)}(\xi,\eta)| - 1)\}, \qquad (2.19)
\end{aligned}
$$

and

$$
\mathcal{F}_4^{(k)}(\xi,\eta)
$$

$$
= -\frac{B^{(k)}(\xi,\eta)\ell^{(k)}}{2A^{(k)}}\mathcal{F}_2^{(k)}(\xi,\eta)
$$

$$
+\frac{[\ell^{(k)}]^2}{4\pi A^{(k)}}[n_x^{(k)}(x^{(k)}-\xi)+n_y^{(k)}(y^{(k)}-\eta)]
$$

$$
\times [\ln|A^{(k)} + B^{(k)}(\xi,\eta) + E^{(k)}(\xi,\eta)| - \ln|E^{(k)}(\xi,\eta)|].
$$

$$
(2.20)
$$

If $(\xi, \eta) \in C^{(k)}$ the integrals in Eq. (2.18) are improper but the formulae in Eqs. (2.19) and (2.20) are still valid provided that (ξ, η) is not given by any one of the two endpoints of $C^{(k)}$. Note that if $4A^{(k)}E^{(k)}(\xi, \eta) - [B^{(k)}(\xi, \eta)]^2 = 0$ then the formula for $\mathcal{F}_4^{(k)}(\xi, \eta)$ in Eq. (2.20) simplifies to a mere $\mathcal{F}_4^{(k)}(\xi, \eta) = 0$.

2.4 Implementation on Computer

The FORTRAN 77 subroutine CPF in Chapter 1 computes and returns the values of $\pi\mathcal{F}_1^{(k)}(\xi, \eta)$ and $\pi\mathcal{F}_2^{(k)}(\xi, \eta)$. We now add to it the computation of $\pi\mathcal{F}_3^{(k)}(\xi, \eta)$ and $\pi\mathcal{F}_4^{(k)}(\xi, \eta)$ in a new subroutine called DPF. DPF takes the values of ξ, η, $x^{(k)}$, $y^{(k)}$, $n_x^{(k)}$, $n_y^{(k)}$ and $\ell^{(k)}$ in the real variables xi, eta, xk, yk, nkx, nky and L respectively and gives the values of $\pi\mathcal{F}_1^{(k)}(\xi, \eta)$, $\pi\mathcal{F}_2^{(k)}(\xi, \eta)$, $\pi\mathcal{F}_3^{(k)}(\xi, \eta)$ and $\pi\mathcal{F}_4^{(k)}(\xi, \eta)$ (computed using Eqs. (2.12), (2.13), (2.14), (2.15), (2.19) and (2.20)) in the real variables PF1, PF2, PF3 and PF4 respectively.

The subroutine DPF is listed below.

```
      subroutine DPF(xi,eta,xk,yk,nkx,nky,L,
     & PF1,PF2,PF3,PF4)

      double precision xi,eta,xk,yk,nkx,nky,L,
     & PF1,PF2,PF3,PF4,A,B,E,D,BA,EA

      A=L**2d0
      B=2d0*L*(-nky*(xk-xi)+nkx*(yk-eta))
      E=(xk-xi)**2d0+(yk-eta)**2d0
      D=dsqrt(dabs(4d0*A*E-B**2d0))
      BA=B/A
      EA=E/A

      if (D.lt.0.0000000001d0) then
      PF1=0.5d0*L*(dlog(L)
     & +(1d0+0.5d0*BA)*dlog(dabs(1d0+0.5d0*BA))
     & -0.5d0*BA*dlog(dabs(0.5d0*BA))-1d0)
```

```
      PF2=0d0
      PF4=0d0
      else
      PF1=0.25d0*L*(2d0*(dlog(L)-1d0)
     & -0.5d0*BA*dlog(dabs(EA))
     & +(1d0+0.5d0*BA)*dlog(dabs(1d0+BA+EA))
     & +(D/A)*(datan((2d0*A+B)/D)-datan(B/D)))
      PF2=L*(nkx*(xk-xi)+nky*(yk-eta))/D
     & *(datan((2d0*A+B)/D)-datan(B/D))
      PF4=-0.5d0*BA*L*PF2
     & +(0.25d0*L*L/A)*(nkx*(xk-xi)+nky*(yk-eta))
     & *(dlog(dabs(A+B+E))-dlog(dabs(E)))
      endif

      PF3=-0.5d0*BA*L*PF1+(L*L/(8d0*A))
     & *((A+B+E)*(dlog(dabs(A+B+E))-1d0)
     & -E*(dlog(dabs(E))-1d0))

      return
      end
```

We now create the subroutine DLELAP1 which sets up the system of linear algebraic equations in Eq. (2.10) and solve it to determine $\widehat{\phi}^{(k)}$, $\widehat{\phi}^{(N+k)}$ $\widehat{p}^{(k)}$ and $\widehat{p}^{(N+k)}$ ($k = 1, 2, \cdots, N$). The values of $\widehat{\phi}^{(m)}$ and $\widehat{p}^{(m)}$ ($m = 1, 2, \cdots, 2N$) are returned in the arrays phi(1:2*N) and dphi(1:2*N) respectively. The input parameters required by DLELAP1 are N (the integer N), the constant τ in in Eqs. (2.5) and (2.6) ($0 < \tau < 1/2$) (the real variable tau), ($\xi^{(m)}, \eta^{(m)}$) ($m = 1, 2, \cdots, 2N$) (stored inside the real arrays xm(1:2*N) and ym(1:2*N)), the boundary points ($x^{(k)}, y^{(k)}$) (in the real arrays xb(1:N+1) and yb(1:N+1)), the unit normal vector $[n_x^{(k)}, n_y^{(k)}]$ (in the real arrays nx(1:N) and ny(1:N)), the lengths of the elements (in the real array lg(1:N)), the types of boundary conditions (in the integer array BCT(1:N)) (as explained on page 39) and the specified values of either $\widehat{\phi}^{(m)}$ or $\widehat{p}^{(m)}$ ($m = 1, 2, \cdots, 2N$) (in the real arrays BCV(1:2*N)).

As an example, if ϕ is known on $C^{(2)}$ and if $N = 10$, then the values of ϕ at the two points $(\xi^{(2)}, \eta^{(2)})$ and $(\xi^{(12)}, \eta^{(12)})$ (on $C^{(2)}$) (stored in the real variables xm(2), ym(2), xm(12) and ym(12)) are given in the real variables BCV(2) and BCV(12) respectively.

```
      subroutine DLELAP1(N,tau,xm,ym,xb,yb,
     & nx,ny,lg,BCT,BCV,phi,dphi)

      integer m,k,N,BCT(1000)

      double precision xm(1000),ym(1000),
     & xb(1000),yb(1000),nx(1000),ny(1000),lg(1000),
     & BCV(1000),A(1000,1000),B(1000),pi,PF1,PF2,
     & delkm,delnkm,phi(1000),dphi(1000),F1,F2,
     & Z(1000),PF3,PF4,F3,F4,tau

      pi=4d0*datan(1d0)

      do 10 m=1,2*N
      B(m)=0d0
      do 5 k=1,N
      call DPF(xm(m),ym(m),xb(k),yb(k),nx(k),ny(k),
     & lg(k),PF1,PF2,PF3,PF4)
      F1=PF1/pi
      F2=PF2/pi
      F3=PF3/pi
      F4=PF4/pi
      if (k.eq.m) then
      delkm=1d0
      else
      delkm=0d0
      endif
      if (k.eq.(m-N)) then
      delnkm=1d0
      else
      delnkm=0d0
```

```fortran
      endif
      if (BCT(k).eq.0) then
      A(m,k)=((1d0-tau)*F1-F3/lg(k))/(2d0*tau-1d0)
      A(m,N+k)=(-tau*F1+F3/lg(k))/(2d0*tau-1d0)
      B(m)=B(m)-((-(1d0-tau)*F2+F4/lg(k))/
     & (2d0*tau-1d0)-0.5d0*delkm)*BCV(k)
     & -((tau*F2-F4/lg(k))/
     & (2d0*tau-1d0)-0.5d0*delnkm)*BCV(N+k)
      else
      A(m,k)=(-(1d0-tau)*F2+F4/lg(k))
     & /(2d0*tau-1d0)-0.5d0*delkm
      A(m,N+k)=(tau*F2-F4/lg(k))
     & /(2d0*tau-1d0)-0.5d0*delnkm
      B(m)=B(m)-(((1d0-tau)*F1-F3/lg(k))*BCV(k)
     & +(-tau*F1+F3/lg(k))*BCV(N+k))/(2d0*tau-1d0)
      endif
5     continue
10    continue

      call solver(A,B,2*N,1,Z)

      do 15 m=1,N
      if (BCT(m).eq.0) then
      phi(m)=BCV(m)
      phi(N+m)=BCV(N+m)
      dphi(m)=Z(m)
      dphi(N+m)=Z(N+m)
      else
      phi(m)=Z(m)
      phi(N+m)=Z(N+m)
      dphi(m)=BCV(m)
      dphi(N+m)=BCV(N+m)
      endif
15    continue

      return
      end
```

The subroutine SOLVER used by DLELAP1 for solving the system of linear algebraic equations is listed on page 30.

The output values stored in the real arrays phi(1:2*N) and dphi(1:2*N), as returned by DLELAP1, are used in the subroutine DLELAP2 to compute ϕ at any input point (ξ, η) (whose coordinates are stored in the real variables xi and eta). DLELAP2 uses Eq. (2.7) with $\lambda(\xi, \eta) = 1$ to compute ϕ and returns the desired value in pint.

```
      subroutine DLELAP2(N,tau,xi,eta,xb,yb,
     & nx,ny,lg,phi,dphi,pint)

      integer N,i

      double precision xi,eta,xb(1000),yb(1000),
     & nx(1000),ny(1000),lg(1000),phi(1000),
     & dphi(1000),pint,sum,pi,PF1,PF2,PF3,PF4,tau

      pi=4d0*datan(1d0)
      sum=0d0

      do 10 i=1,N
      call DPF(xi,eta,xb(i),yb(i),nx(i),ny(i),lg(i),
     &   PF1,PF2,PF3,PF4)
      sum=sum+(phi(i)*(-(1d0-tau)*lg(i)*PF2+PF4)
     & +phi(N+i)*(tau*lg(i)*PF2-PF4)
     & -dphi(i)*(-(1d0-tau)*lg(i)*PF1+PF3)
     & -dphi(N+i)*(tau*lg(i)*PF1-PF3))/lg(i)
 10   continue

      pint=sum/(pi*(2d0*tau-1d0))

      return
      end
```

Note that the subroutines `DLELAP1` and `DLELAP2` for the discontinuous linear elements as listed above allow for a maximum of 1000 unknowns in the system of linear algebraic equations in Eq. (2.10). Thus, in using these subroutines, we may discretize the boundary of the solution domain into not more than 500 boundary elements, as each element contains 2 unknowns.

2.5   Numerical Examples

Example 2.1

Consider again the boundary value problem in Example 1.2 which requires us to solve the Laplace's equation in Eq. (2.1) in the region $1 < x^2 + y^2 < 4$, $x > 0$, $y > 0$, subject to

$$\frac{\partial \phi}{\partial n} = 0 \text{ on the straight side } x = 0, \ 1 < y < 2,$$

$$\frac{\partial \phi}{\partial n} = 0 \text{ on the straight side } y = 0, \ 1 < x < 2,$$

$$\phi = \cos(4\arctan(\frac{y}{x})) \text{ on the arc } x^2 + y^2 = 1, \ x > 0, \ y > 0,$$

$$\phi = 3\cos(4\arctan(\frac{y}{x})) \text{ on the arc } x^2 + y^2 = 4, \ x > 0, \ y > 0.$$

A geometrical sketch of the problem is given in Figure 1.6 on page 45.

We solve the problem using the boundary element method with discontinuous boundary elements, that is, through the use of the subroutines `DLELAP1` and `DLELAP2`. The real arrays `xb(1:N+1)` and `yb(1:N+1)` containing the input boundary points $(x^{(k)}, y^{(k)})$ ($k = 1, 2, \cdots, N$) can be generated as in Example 1.2 using the code given on page 46.

For discontinuous linear elements, the real arrays `xm(1:2*N)` and `ym(1:2*N)` contain the coordinates of the points $(\xi^{(k)}, \eta^{(k)})$ and $(\xi^{(N+k)}, \eta^{(N+k)})$ which lie on $C^{(k)}$ at a distance of $\tau \ell^{(k)}$ from the boundary points $(x^{(k)}, y^{(k)})$ and $(x^{(k+1)}, y^{(k+1)})$. With the input value of τ given in the real variable `tau`, the code for setting up the arrays `xm(1:2*N)` and `ym(1:2*N)` is as follows.

```
      do 20 i=1,N
      xm(i)=xb(i)+tau*(xb(i+1)-xb(i))
      ym(i)=yb(i)+tau*(yb(i+1)-yb(i))
      xm(N+i)=xb(i)+(1d0-tau)*(xb(i+1)-xb(i))
      ym(N+i)=yb(i)+(1d0-tau)*(yb(i+1)-yb(i))
                    ⋮
                    ⋮
   20 continue
```

The integer array BCT(1:N) and the real arrays BCV(1:2*N)
which give the input data for the boundary conditions on the
elements may be generated using the following code.

```
      do 30 i=1,N
      if ((i.le.N0).or.((i.gt.(9*N0))
   &  .and.(i.le.(10*N0)))) then
      BCT(i)=1
      BCV(i)=0d0
      BCV(N+i)=0d0
      else if ((i.gt.N0).and.(i.le.(9*N0))) then
      BCT(i)=0
      BCV(i)=3d0*dcos(4d0*datan(ym(i)/xm(i)))
      BCV(N+i)=3d0*dcos(4d0*datan(ym(N+i)/xm(N+i)))
      else
      BCT(i)=0
      BCV(i)=dcos(4d0*datan(ym(i)/xm(i)))
      BCV(N+i)=dcos(4d0*datan(ym(N+i)/xm(N+i)))
      endif
   30 continue
```

A complete program EXP2PT1 for solving the boundary value
problem through the use of discontinuous linear elements is as
listed below.

```
      program EX2PT1

      integer N0,BCT(1000),N,i,ians
```

```fortran
      double precision xb(1000),yb(1000),xm(1000),
     & ym(1000),nx(1000),ny(1000),lg(1000),BCV(1000),
     & phi(1000),dphi(1000),pint,dl,xi,eta,pi,tau

      print*,'Enter integer N0 (<42):'
      read*,N0
      N=12*N0
      print*,'Enter parameter tau (0<tau<1/2):'
      read*,tau

      pi=4d0*datan(1d0)

      do 10 i=1,8*N0
      dl=pi/dfloat(16*N0)
      xb(i+N0)=2d0*dcos(dfloat(i-1)*dl)
      yb(i+N0)=2d0*dsin(dfloat(i-1)*dl)
      if (i.le.N0) then
      dl=1d0/dfloat(N0)
      xb(i)=1d0+dfloat(i-1)*dl
      yb(i)=0d0
      xb(i+9*N0)=0d0
      yb(i+9*N0)=2d0-dfloat(i-1)*dl
      endif
      if (i.le.(2*N0)) then
      dl=pi/dfloat(4*N0)
      xb(i+10*N0)=dsin(dfloat(i-1)*dl)
      yb(i+10*N0)=dcos(dfloat(i-1)*dl)
      endif
10    continue
      xb(N+1)=xb(1)
      yb(N+1)=yb(1)

      do 20 i=1,N
      xm(i)=xb(i)+tau*(xb(i+1)-xb(i))
      ym(i)=yb(i)+tau*(yb(i+1)-yb(i))
      xm(N+i)=xb(i)+(1d0-tau)*(xb(i+1)-xb(i))
```

```fortran
      ym(N+i)=yb(i)+(1d0-tau)*(yb(i+1)-yb(i))
      lg(i)=dsqrt((xb(i+1)
     & -xb(i))**2d0+(yb(i+1)-yb(i))**2d0)
      nx(i)=(yb(i+1)-yb(i))/lg(i)
      ny(i)=(xb(i)-xb(i+1))/lg(i)
20 continue

      do 30 i=1,N
      if ((i.le.N0).or.((i.gt.(9*N0))
     & .and.(i.le.(10*N0)))) then
      BCT(i)=1
      BCV(i)=0d0
      BCV(N+i)=0d0
      else if ((i.gt.N0).and.(i.le.(9*N0))) then
      BCT(i)=0
      BCV(i)=3d0*dcos(4d0*datan(ym(i)/xm(i)))
      BCV(N+i)=3d0*dcos(4d0*datan(ym(N+i)/xm(N+i)))
      else
      BCT(i)=0
      BCV(i)=dcos(4d0*datan(ym(i)/xm(i)))
      BCV(N+i)=dcos(4d0*datan(ym(N+i)/xm(N+i)))
      endif
30 continue

      call DLELAP1(N,tau,xm,ym,xb,yb,nx,ny,
     & lg,BCT,BCV,phi,dphi)

50 print*,'Enter coordinates xi and eta of
     & an interior point:'
      read*,xi,eta

      call DLELAP2(N,tau,xi,eta,xb,yb,nx,ny,
     & lg,phi,dphi,pint)

      dl=(xi**2d0+eta**2d0)**2d0
      dl=((16d0/85d0)*(dl-1d0/dl)
     &   -(16d0/255d0)*(dl/16d0-16d0/dl))
```

```
&   *dcos(4d0*datan(eta/xi))

    write(*,60)pint,dl,xi,eta
60 format('Numerical and exact values are:',
&   F14.6,' and',F14.6,' respectively at
&   (',F14.6,',',F14.6,')')

    print*,'To continue with another point enter 1:'
    read*,ians

    if (ians.eq.1) goto 50

    end
```

Table 2.1

(ξ,η)	Constant elements $N = 240$	Discontinuous linear elements $N = 120$
$(1.082532, 0.625000)$	-0.392546	-0.392098
$(0.875000, 1.515544)$	-0.908254	-0.908057
$(1.060660, 1.060660)$	-1.094489	-1.094335
$(1.099998, 0.001920)$	0.824548	0.821265
$(1.010000, 0.000176)$	0.960174	0.972723

(ξ,η)	Discontinuous linear elements $N = 240$	Exact
$(1.082532, 0.625000)$	-0.392055	-0.392045
$(0.875000, 1.515544)$	-0.907875	-0.907816
$(1.060660, 1.060660)$	-1.094240	-1.094211
$(1.099998, 0.001920)$	0.825730	0.826958
$(1.010000, 0.000176)$	0.975461	0.975656

The numerical values of ϕ at selected points in the interior of the solution domain obtained by the program **EX2PT1** using 120 and 240 boundary elements with $\tau = 0.25$ are compared with those calculated using 240 constant elements as well as the exact solution in Table 2.1. From the table, it is apparent that the overall performance of the discontinuous linear elements is better than the constant ones for the same number of either unknowns or elements. (Note that, for the same number of boundary elements, the number of unknowns involved in the formulation for the discontinuous linear elements is twice that for the constant elements.)

For a more extensive comparison, we compute ϕ at 10000 randomly selected interior points using both the constant and the discontinuous linear elements with $\tau = 0.25$. Discretizing the boundary of the solution domain into 120 boundary elements, we find that the average percentage error of the numerical values is 0.1442% for the discontinuous linear elements. When the calculation is repeated using 240 boundary elements, the average percentage errors for the constant and the discontinuous linear elements are 0.5359% and 0.03488% respectively. On the whole, the discontinuous linear elements outperform the constant ones for the same number of either unknowns or elements. Note that the average percentage error of the numerical values of ϕ at the 10000 points is calculated in accordance with the formula

$$
\text{'average percentage error'}
$$
$$
= \frac{1}{10000} \sum_{i=1}^{10000} \left| \frac{\phi_{\text{numerical}}^{(i)} - \phi_{\text{exact}}^{(i)}}{\phi_{\text{exact}}^{(i)}} \right| \times 100\%
$$

where $\phi_{\text{numerical}}^{(i)}$ and $\phi_{\text{exact}}^{(i)}$ are respectively the numerical and exact values of ϕ at the i-th point. Note that $\phi \neq 0$ at the selected points.

Lastly, we examine the influence of the parameter τ on the accuracy of the numerical values of ϕ obtained by using the discontinuous elements. The average percentage error of the numerical values at the 10000 randomly selected points, as obtained by using 240 elements, is plotted against τ ($0 < \tau < 1/2$) in Figure 2.2. It appears that the value of τ giving the lowest average error is approximately 0.21 for the particular boundary value problem here. With $\tau = 0.21$, the average percentage error is at a low 0.009721%. Further investigations suggest that the optimal value of τ giving the lowest error may vary slightly when a different number of boundary elements is used.

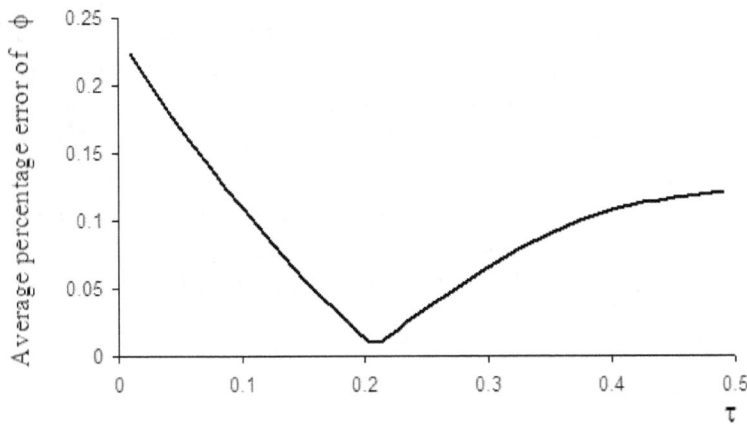

Figure 2.2

Example 2.2

We continue to compare the accuracy of the constant and the discontinuous linear elements by solving another boundary value problem governed by Eq. (2.1). The solution domain is the square region $0 < x < 1$, $0 < y < 1$, and the boundary conditions are

$$\phi(x,0) = 1 \text{ and } \phi(x,1) = 0 \text{ for } 0 < x < 1,$$
$$\phi(0,y) = 0 \text{ and } \phi(1,y) = 0 \text{ for } 0 < y < 1.$$

The exact solution of the boundary value problem here is given by[1]

---

[1]For the derivation of this solution, refer to the text book *Elementary Differential Equations with Boundary Value Problems* by CH Edwards Jr and DE Penney (Prentice-Hall, 1993).

$$\phi(x,y) \;=\; \frac{4}{\pi}\sum_{n=0}^{\infty}\frac{[1-\exp(-2\pi[2n+1][1-y])]}{(2n+1)[1-\exp(-2\pi[2n+1])]}$$
$$\times\,\exp(-\pi[2n+1]y)\sin(\pi[2n+1]x).$$

To solve the problem numerically using the constant and the discontinuous linear elements, each side of the square is discretized into N_0 equal length boundary elements. As in Example 2.1, for both the constant and the discontinuous linear elements, we compute ϕ numerically at 10000 randomly selected points in the interior the solution domain and calculate the average percentage error of the numerical values. For this example, the value of τ giving the smallest average percentage error for the discontinuous linear elements is found to vary between 0.25 and 0.30 depending on the number of elements used.

Table 2.2

Number of elements	Constants elements	Discontinuous linear elements $\tau = 0.05$	Discontinuous linear elements $\tau = 0.25$
20	36.9%	50.8%	5.6%
40	10.1%	9.4%	2.0%
80	1.1%	0.7%	0.1%
160	0.2%	0.04%	0.01%
320	0.07%	0.005%	0.0009%

Table 2.2 compares the average percentage errors for the constant elements and the discontinuous linear elements with $\tau = 0.05$ and $\tau = 0.25$. For $\tau = 0.25$, the accuracy of the discontinuous linear elements is clearly better than that of the constant elements. For $\tau = 0.05$, discontinuous linear elements perform poorly-worst than the constant elements-when the boundary is discretized into very few elements. Nevertheless, the discontinuous linear elements with $\tau = 0.05$ gain superiority in accuracy over the constant elements when the number of elements increase.

2.6 Summary and Discussion

A boundary element solution with discontinuous linear elements is obtained for the interior boundary value problem governed by the two-dimensional Laplace's equation. The discontinuous linear elements are defined by Eqs. (2.5) and (2.6). On each boundary elements, there are 2 unknown constants. Thus, if the boundary of the solution domain is discretized into N straight line segments, the boundary element formulation of the boundary value problem involves a system of $2N$ linear algebraic equations containing $2N$ unknowns.

The accuracy of the boundary element solution depends on the parameter τ (in Eqs. (2.5) and (2.6)) which lies between 0 and $1/2$. The value of τ giving the best results may vary from one boundary value problem to another. For a given problem, it may also change with the number of boundary elements used. Numerical results obtained here for some specific problems using $\tau = 0.25$ indicate that the overall accuracy of the discontinuous linear elements is better than that of the constant elements for the same number of either boundary elements or unknowns on the boundary.

Further investigations using various other numerical examples suggest that the optimal value of τ is likely to be somewhere between 0.15 and 0.30. If very few elements are used, the accuracy of the numerical solution deteriorates rapidly as τ gets closer to zero. On the other hand, if the discretization of the boundary is sufficiently refined, the numerical solutions obtained using the discontinuous linear elements are found to be superior to those given by the constant elements, even if the value of τ is taken to be close to zero[2]. In general, for the discontinuous linear elements to deliver accurate solutions, one

---

[2]In the article "A comparison of boundary element formulations for steady-state heat conduction problems" in the journal *Engineering Analysis with Boundary Elements* (Volume 25, 2001, pp. 115-128), NS Mera, L Elliot, DB Ingham and D Lesnic reported that the optimal value of τ is highly likely to be between 0.10 and 0.25. The governing partial differential equation in their studies may be reduced to the Laplace equation as a special case.

may recommend that τ be given any value between 0.20 and 0.25.

2.7 Exercises

1. How would you modify Eqs. (2.5), (2.6) and (2.7) if the linear approximations of ϕ and $\partial\phi/\partial n$ across the boundary elements are required to be continuous at the endpoints between the elements? How does your modification affect the system of linear algebraic equations in Eq. (2.9) and the analytical formulae for evaluating $\mathcal{F}_1^{(k)}(\xi,\eta)$, $\mathcal{F}_2^{(k)}(\xi,\eta)$, $\mathcal{F}_3^{(k)}(\xi,\eta)$ and $\mathcal{F}_4^{(k)}(\xi,\eta)$?

2. Modify the program EX2PT1 in Section 2.5 to solve numerically the Laplace's equation given by Eq. (2.1) in the region $x^2 + y^2 < 1$, $x > 0$, $y > 0$, subject to the boundary conditions

$$\phi = 0 \text{ on the horizontal side } y = 0 \text{ for } 0 < x < 1,$$
$$\phi = 1 \text{ on the vertical side } x = 0 \text{ for } 0 < y < 1,$$
$$\frac{\partial\phi}{\partial n} = 0 \text{ on the quarter circle}$$
$$x^2 + y^2 = 1, \ x > 0, \ y > 0.$$

Discretize each of the three parts of the boundary (the two sides and the quarter circle) into N_0 boundary elements so that $N = 3N_0$, that is, so that we have a total of $3N_0$ boundary elements. For various values of N_0, by computing ϕ at 5000 randomly selected points in the interior of the solution domain, find the value of the parameter τ (defining the discontinuous linear elements) that gives the lowest average percentage error. (Note. The exact solution of the boundary value problem here is $\phi = (2/\pi)\arctan(y/x)$.)

3. Modify the program EX2PT1 in Section 2.5 to solve numerically the Laplace's equation given by Eq. (2.1) in

the region $0 < x < 1, 0 < y < 1$, subject to the boundary conditions

$$\phi(0,y) = 0 \text{ and } \phi(1,y) = 0 \text{ for } 0 < y < 1,$$
$$\phi(x,0) = 0 \text{ and } \phi(x,1) = \sin(\pi x) \text{ for } 0 < x < 1.$$

Discretize each side of the square into N_0 equal length boundary elements so that $N = 4N_0$. Compute ϕ at some selected interior points and compare the numerical values obtained with the exact solution $\phi(x,y) = \sin(\pi x) \sinh(\pi y)/\sinh(\pi)$. (Explore with different values of N_0 and τ.)

4. Use the subroutines **DLELAP1** and **DLELAP2** to solve numerically the Poisson's equation

$$\frac{\partial^2 \psi}{\partial x^2} + \frac{\partial^2 \psi}{\partial y^2} = 2y \text{ in the region } 0 < x < 1, \ 0 < y < 1,$$

subject to

$$\psi(0,y) = 0 \text{ and } \psi(1,y) = y \text{ for } 0 < y < 1,$$
$$\psi(x,0) = 0 \text{ and } \psi(x,1) = x^2 + \sin(\pi x) \text{ for } 0 < x < 1.$$

Calculate $\psi(0.25, 0.75)$ numerically. (Note. To use the subroutines **DLELAP1** and **DLELAP2**, change the governing partial differential equation from the Poisson's equation to the Laplace's equation through a suitable substitution.)

5. Modify the program **EX2PT1** in Section 2.5 to solve numerically the Laplace's equation given by Eq. (2.1) in the region $x^2 + y^2 < 1$ given that $\phi = 1$ on the part of the circle in the first quadrant ($x > 0$, $y > 0$) and $\phi = 0$ on the remaining part of the circle. Discretize the circle into N boundary elements of equal length. For various values of N, check your numerical values of ϕ at some interior points against the exact solution given by $\phi(x,y) = \pi/2 + (1/\pi) \arctan([1 - x^2 - y^2]/[(x-1)^2 + (y-1)^2 - 1])$.

If we define $\psi(r,\theta) = \phi(r\cos\theta, r\sin\theta)$, where r and θ are polar coordinates given by $x = r\cos\theta$ and $y = r\sin\theta$, explain how you can use your numerical solution to compute $\partial\psi/\partial\theta$ across every boundary element.

Chapter 3

Two-dimensional Helmholtz Type Equation

3.1   Introduction

Of interest here is the numerical solution of the two-dimensional Helmholtz type equation of the form

$$\frac{\partial^2 \phi}{\partial x^2} + \frac{\partial^2 \phi}{\partial y^2} + \alpha(x,y)\phi = g(x,y), \qquad (3.1)$$

where α and g are suitably given functions of x and y.

If we let $\alpha(x,y) = 0$ and $g(x,y) = 0$ at all points (x,y) in the solution domain, we recover from Eq. (3.1) the two-dimensional Laplace's equation in Chapter 1. Governing partial differential equations of many problems, such those as in acoustics, bioheat transfer and mechanics of nonhomogeneous materials, may be reducible to the form in Eq. (3.1) with α being a non-zero constant or a smoothly varying function of x and y and the function g representing source terms (for example, internal heat generation in the case of a thermal problem or gravitational body force in mechanics).

As in Chapter 1, Eq. (3.1) is to be solved in the two-dimensional region R bounded by a simple closed curve C subject to the boundary conditions

$$\phi = f_1(x, y) \text{ for } (x, y) \in C_1,$$
$$\frac{\partial \phi}{\partial n} = f_2(x, y) \text{ for } (x, y) \in C_2, \qquad (3.2)$$

where f_1 and f_2 are suitably prescribed functions and C_1 and C_2 are non-intersecting curves such that $C_1 \cup C_2 = C$. Refer to Figure 1.1 on page 8 for a geometrical sketch of the problem.

This chapter comprises two independent parts.

In the first part, we consider the special case $\alpha(x, y) = w^2$ (w is a positive real constant) and $g(x, y) = 0$ at all points (x, y) in R, in which Eq. (3.1) is reduced to the two-dimensional homogeneous Helmholtz equation given by

$$\frac{\partial^2 \phi}{\partial x^2} + \frac{\partial^2 \phi}{\partial y^2} + w^2 \phi = 0. \qquad (3.3)$$

For this particular case, the analyses in Sections 1.2, 1.3 and 1.4 of Chapter 1 may be used as a guide to derive the boundary integral equation

$$\lambda(\xi, \eta)\phi(\xi, \eta) = \int_C [\phi(x, y)\frac{\partial}{\partial n}(\Omega(x, y; \xi, \eta))$$
$$-\Omega(x, y; \xi, \eta)\frac{\partial}{\partial n}(\phi(x, y))]ds(x, y), (3.4)$$

where

$$\lambda(\xi, \eta) = \begin{cases} 0 & \text{if } (\xi, \eta) \notin R \cup C, \\ 1/2 & \text{if } (\xi, \eta) \text{ lies on a smooth part of } C, \\ 1 & \text{if } (\xi, \eta) \in R, \end{cases} \qquad (3.5)$$

$\Omega(x, y; \xi, \eta)$ is the fundamental solution of the homogeneous Helmholtz equation as given by

$$\Omega(x, y; \xi, \eta) = \frac{1}{4}Y_0(w\sqrt{(x - \xi)^2 + (y - \eta)^2}), \qquad (3.6)$$

and Y_0 is the zeroth order Bessel function of the second kind. The boundary integral equation in Eq. (3.4) may be discretized

82

as in Chapter 1 to solve numerically Eq. (3.3) in R subject to Eq. (3.2).

The second part of the chapter is concerned with the numerical solution of Eqs. (3.1)-(3.2) for general functions α and g which vary smoothly from point to point in the solution domain. For such a general case, an integral formulation may be obtained for Eq. (3.1) by using the fundamental solution $\Phi(x, y; \xi, \eta)$ of the two-dimensional Laplace's equation. The formulation does not only contain an integral over the boundary C but also a double integral over the entire solution domain R. To avoid having to discretize the solution domain into many tiny cells, the so called dual-reciprocity method is used to convert the double integral approximately into a line integral over C. To do this, the term $g(x, y) - \alpha(x, y)\phi$ which appears in the integrand of the double integral is approximated using appropriate radial basis functions. In applying the integral formulation to set up a system of linear algebraic equations for the numerical solution of Eqs. (3.1)-(3.2), only the boundary C has to be discretized into elements. Nevertheless, the unknowns in the algebraic equations do not involve only yet to be determined values of ϕ or $\partial\phi/\partial n$ on the boundary elements but also those of ϕ at selected collocation points in the interior of R.

3.2 Homogeneous Helmholtz Equation

3.2.1 Fundamental Solution

If $x = r\cos\theta$ and $y = r\sin\theta$ and ϕ is independent of θ such that $\phi = \psi(r)$, we find that Eq. (3.3) may be written as

$$r\psi''(r) + \psi'(r) + w^2 r\psi(r) = 0. \tag{3.7}$$

Using the method of Frobenius[1], we find that Eq. (3.7)

---

[1]For details, one may refer to the text book *Differential Equations for Mathematics, Science, and Engineering* by PW Davis (Prentice-Hall International Editions).

admits the general solution

$$\psi(r) = AY_0(wr) + BJ_0(wr), \qquad (3.8)$$

where A and B are arbitrary constants and J_0 and Y_0 are the zeroth order Bessel functions of the first and second kinds respectively. Note once again that w is a positive real number.

The Bessel functions J_0 and Y_0 are given by

$$
\begin{aligned}
J_0(x) &= \sum_{m=0}^{\infty} \frac{(-1)^m x^{2m}}{4^m (m!)^2}, \\
Y_0(x) &= \frac{2}{\pi}(\ln(\frac{x}{2}) + \gamma)J_0(x) - \frac{2}{\pi}\sum_{m=1}^{\infty} \frac{(-1)^m x^{2m}}{4^m (m!)^2} \sum_{k=1}^{m} \frac{1}{k},
\end{aligned}
$$
$$(3.9)$$

where γ is the Euler constant defined by

$$\gamma = \lim_{n\to\infty}(-\ln(n) + 1 + \frac{1}{2} + \frac{1}{3} + \cdots + \frac{1}{n}). \qquad (3.10)$$

From (3.9), it is obvious that to ensure that $\psi(r)$ behaves like $(2\pi)^{-1}\ln(r)$ (that is, like the fundamental solution of the two-dimensional Laplace's equation in Chapter 1) for small r, we may take $A = 1/4$ and $B = 0$, that is, we take

$$\psi(r) = \frac{1}{4}Y_0(wr). \qquad (3.11)$$

More generally, a particular solution of Eq. (3.3) is given by

$$\phi(x,y) = \frac{1}{4}Y_0(w\sqrt{(x-\xi)^2 + (y-\eta)^2}) \text{ for } (x,y) \neq (\xi,\eta). \qquad (3.12)$$

We denote this particular solution by $\Omega(x,y;\xi,\eta)$ as in Eq. (3.6) and refer to it as the fundamental solution of Eq. (3.3).

3.2.2 Boundary Integral Solution

If we let ϕ_1 and ϕ_2 be two solutions of Eq. (3.3) in the region R bounded by the simple closed curve C, that is,

$$\frac{\partial^2 \phi_1}{\partial x^2} + \frac{\partial^2 \phi_1}{\partial y^2} + w^2 \phi_1 = 0,$$

$$\frac{\partial^2 \phi_2}{\partial x^2} + \frac{\partial^2 \phi_2}{\partial y^2} + w^2 \phi_2 = 0, \qquad (3.13)$$

we may proceed as in Section 1.3 (page 11, Chapter 1) to derive the reciprocal relation

$$\int_C (\phi_2 \frac{\partial \phi_1}{\partial n} - \phi_1 \frac{\partial \phi_2}{\partial n}) ds(x,y) = 0. \qquad (3.14)$$

Guided by the analysis in Section 1.4 (page 12, Chapter 1), we may use Eqs. (3.9) and (3.14) by letting $\phi_1 = \Omega(x, y; \xi, \eta)$ (as defined in Eq. (3.6)) and $\phi_2 = \phi$ (the solution of the boundary value problem defined by Eqs. (3.1)-(3.2)) to derive the boundary integral solution in Eq. (3.4).

3.2.3 Numerical Procedure

The boundary integral solution in Eq. (3.4) may be applied as in Section 1.5 to obtain a simple boundary element procedure for solving Eqs. (3.1)-(3.2) numerically.

If the boundary C is discretized into N straight line segments $C^{(1)}, C^{(2)}, \cdots, C^{(N-1)}$ and $C^{(N)}$ and ϕ and $\partial \phi / \partial n$ are respectively approximated as constants $\overline{\phi}^{(k)}$ and $\overline{p}^{(k)}$ over the line segment $C^{(k)}$, we find that Eq. (3.4) can be approximately written as

$$\lambda(\xi, \eta) \phi(\xi, \eta) = \sum_{k=1}^{N} \{\overline{\phi}^{(k)} \mathcal{G}_2^{(k)}(\xi, \eta) - \overline{p}^{(k)} \mathcal{G}_1^{(k)}(\xi, \eta)\}, \qquad (3.15)$$

where

$$\mathcal{G}_1^{(k)}(\xi,\eta) = \int\limits_{C^{(k)}} \Omega(x,y;\xi,\eta)ds(x,y),$$

$$\mathcal{G}_2^{(k)}(\xi,\eta) = \int\limits_{C^{(k)}} \frac{\partial}{\partial n}[\Omega(x,y;\xi,\eta)]ds(x,y). \qquad (3.16)$$

If we let (ξ,η) in Eq. (3.15) be given in turn by the midpoints of $C^{(1)}$, $C^{(2)}$, $\cdots$, $C^{(N-1)}$ and $C^{(N)}$, we obtain

$$\frac{1}{2}\overline{\phi}^{(m)} = \sum_{k=1}^{N}\{\overline{\phi}^{(k)}\mathcal{G}_2^{(k)}(\overline{x}^{(m)},\overline{y}^{(m)}) - \overline{p}^{(k)}\mathcal{G}_1^{(k)}(\overline{x}^{(m)},\overline{y}^{(m)})\}$$

$$\text{for } m = 1,2,\cdots,N, \qquad (3.17)$$

where $(\overline{x}^{(m)},\overline{y}^{(m)})$ is the midpoint of $C^{(m)}$.

Since either $\overline{\phi}^{(k)}$ or $\overline{p}^{(k)}$ is known for a given k, Eq. (3.17) constitutes a system of N linear algebraic equations containing the N unknowns. It may be rewritten as

$$\sum_{k=1}^{N} a^{(mk)}z^{(k)} = \sum_{k=1}^{N} b^{(mk)} \text{ for } m = 1,2,\cdots,N, \qquad (3.18)$$

where $a^{(mk)}$, $b^{(mk)}$ and $z^{(k)}$ are defined by

$$a^{(mk)} = \begin{cases} -\mathcal{G}_1^{(k)}(\overline{x}^{(m)},\overline{y}^{(m)}) \text{ if } \phi \text{ is specified over } C^{(k)}, \\ \mathcal{G}_2^{(k)}(\overline{x}^{(m)},\overline{y}^{(m)}) - \frac{1}{2}\delta^{(mk)} \\ \qquad \text{if } \partial\phi/\partial n \text{ is specified over } C^{(k)}, \end{cases}$$

$$b^{(mk)} = \begin{cases} \overline{\phi}^{(k)}(-\mathcal{G}_2^{(k)}(\overline{x}^{(m)},\overline{y}^{(m)}) + \frac{1}{2}\delta^{(mk)}) \\ \qquad \text{if } \phi \text{ is specified over } C^{(k)}, \\ \overline{p}^{(k)}\mathcal{G}_1^{(k)}(\overline{x}^{(m)},\overline{y}^{(m)}) \\ \qquad \text{if } \partial\phi/\partial n \text{ is specified over } C^{(k)}, \end{cases}$$

$$\delta^{(mk)} = \begin{cases} 0 \text{ if } m \neq k, \\ 1 \text{ if } m = k, \end{cases}$$

$$z^{(k)} = \begin{cases} \overline{p}^{(k)} \text{ if } \phi \text{ is specified over } C^{(k)}, \\ \overline{\phi}^{(k)} \text{ if } \partial\phi/\partial n \text{ is specified over } C^{(k)}. \end{cases} \qquad (3.19)$$

Note that $z^{(1)}$, $z^{(2)}$, $\cdots$, $z^{(N-1)}$ and $z^{(N)}$ are the N unknown constants on the right hand side of Eq. (3.17), while $a^{(mk)}$ and $b^{(mk)}$ are known coefficients.

Once Eq. (3.18) is solved for its unknowns, $\overline{\phi}^{(k)}$ and $\overline{p}^{(k)}$ are known for $k = 1, 2, \cdots, N$, and Eq. (3.15) with $\lambda(\xi, \eta) = 1$ gives an explicit formula for computing ϕ in the interior of R, that is,

$$\phi(\xi, \eta) \simeq \sum_{k=1}^{N} \{\overline{\phi}^{(k)} \mathcal{G}_2^{(k)}(\xi, \eta) - \overline{p}^{(k)} \mathcal{G}_1^{(k)}(\xi, \eta)\} \quad \text{for } (\xi, \eta) \in R.$$

$$(3.20)$$

Following closely the analysis in Section 1.6 (page 22, Chapter 1), we find that $\mathcal{G}_1^{(k)}(\xi, \eta)$ and $\mathcal{G}_2^{(k)}(\xi, \eta)$ can be written as

$$\mathcal{G}_1^{(k)}(\xi, \eta)$$
$$= \frac{\ell^{(k)}}{4} \int_0^1 Y_0(w\sqrt{A^{(k)}t^2 + B^{(k)}(\xi, \eta)t + E^{(k)}(\xi, \eta)})dt,$$

$$\mathcal{G}_2^{(k)}(\xi, \eta)$$
$$= -\frac{\ell^{(k)}}{4} \int_0^1 \frac{n_x^{(k)}(x^{(k)} - \xi) + n_y^{(k)}(y^{(k)} - \eta)}{\sqrt{A^{(k)}t^2 + B^{(k)}(\xi, \eta)t + E^{(k)}(\xi, \eta)}}$$
$$\times wY_1(w\sqrt{A^{(k)}t^2 + B^{(k)}(\xi, \eta)t + E^{(k)}(\xi, \eta)})dt,$$

$$(3.21)$$

where $\ell^{(k)}$ is the length of $C^{(k)}$, $(x^{(k)}, y^{(k)})$ is the starting point of the element $C^{(k)}$, the boundary points $(x^{(1)}, y^{(1)})$, $(x^{(2)}, y^{(2)})$, $\cdots$, $(x^{(N-1)}, y^{(N-1)})$ and $(x^{(N)}, y^{(N)})$ are arranged following the counter clockwise direction on C, $[n_x^{(k)}, n_y^{(k)}]$ is the unit normal vector to $C^{(k)}$ pointing away from R, Y_1 is the first order Bessel function of the second kind, and

$$A^{(k)} = [\ell^{(k)}]^2,$$
$$B^{(k)}(\xi, \eta) = [-n_y^{(k)}(x^{(k)} - \xi) + (y^{(k)} - \eta)n_x^{(k)}](2\ell^{(k)}),$$
$$E^{(k)}(\xi, \eta) = (x^{(k)} - \xi)^2 + (y^{(k)} - \eta)^2.$$

$$(3.22)$$

Note that we make use of the formula $Y_0'(x) = -Y_1(x)$ in the derivation of Eq. (3.21).

If $4A^{(k)}E^{(k)}(\xi, \eta) - [B^{(k)}(\xi, \eta)]^2 = 0$ then $\mathcal{G}_2^{(k)}(\xi, \eta) = 0$. In general, it may be difficult (if not impossible) to find analytical formulae for $\mathcal{G}_1^{(k)}(\xi, \eta)$ and $\mathcal{G}_2^{(k)}(\xi, \eta)$. Thus, we resort to numerical integration for computing $\mathcal{G}_1^{(k)}(\xi, \eta)$ and $\mathcal{G}_2^{(k)}(\xi, \eta)$ in Eq. (3.21). For the numerical integration, we use the Gaussian quadrature formula to obtain the approximation

$$
\int_0^1 f(t)dt = \int_0^{1/2} f(t)dt + \int_{1/2}^1 f(t)dt
$$

$$
\simeq \frac{1}{4}\sum_{i=1}^{n} w_i f(\frac{1}{4}x_i + \frac{1}{4}) + \frac{1}{4}\sum_{i=1}^{n} w_i f(\frac{1}{4}x_i + \frac{3}{4}),
$$

(3.23)

where the Gaussian weights w_i and the corresponding abscissas x_i can be found in the literature[2].

If $(\xi, \eta) \notin C^{(k)}$, the integrals defining $\mathcal{G}_1^{(k)}(\xi, \eta)$ and $\mathcal{G}_2^{(k)}(\xi, \eta)$ are proper and the use of Eq. (3.23) for computing them numerically does not pose any major difficulty. If (ξ, η) is the midpoint of $C^{(k)}$, the integral defining $\mathcal{G}_1^{(k)}(\xi, \eta)$ in Eq. (3.21) is improper, with its integrand containing a weak logarithmic singularity at $t = 1/2$. We may ignore this weak singularity in the Gaussian quadrature formula. In Eq. (3.23), the integrand is not evaluated at $t = 1/2$. The numerical integration may have to be refined if the magnitude of the coefficient w in Eq. (3.3) exceeds a certain value.

3.2.4　Implementation on Computer

The calculation of $\mathcal{G}_1^{(k)}(\xi, \eta)$ and $\mathcal{G}_2^{(k)}(\xi, \eta)$ requires us to evaluate the Bessel functions $Y_0(x)$ and $Y_1(x)$. Accurate elemen-

---

[2]See, for example, M Abramowitz and IA Stegun, *Handbook of Mathematical Functions* (Dover, 1970). The weights w_i and the abscissas x_i depend on the integer n in the formulae. In the handbook, w_i and x_i are tabulated for several different values of n.

tary formulae for computing these functions are available. The FORTRAN functions[3] BY0 and BY1 below evaluate $Y_0(x)$ and $Y_1(x)$ respectively.

```
function BY0(x)

double precision x,y,BY0,BJ0,f0,th0

if (x.le.0) then
print*,'Invalid argument in calculation
& of Y0(x)'
goto 20
endif

if (x.lt.3d0) then
y=(x/3d0)**2d0
BJ0=1d0-2.2499997d0*y+1.2656208d0*(y**2d0)
& -0.3163866d0*(y**3d0)+0.0444479d0*(y**4d0)
& -0.0039444d0*(y**5d0)+0.00021d0*(y**6d0)
BY0=0.63661977d0*dlog(0.5d0*x)*BJ0
& +0.36746691d0
& +0.60559366d0*y-0.74350384d0*(y**2d0)
& +0.25300117d0*(y**3d0)-0.04261214d0*(y**4d0)
& +0.00427916d0*(y**5d0)-0.00024846d0*(y**6d0)
else
y=3d0/x
f0=0.79788456d0-0.00000077d0*y
& -0.00552740d0*(y**2d0)
& -0.00009512d0*(y**3d0)+0.00137237d0*(y**4d0)
& -0.00072805*(y**5d0)+0.00014476d0*(y**6d0)
th0=x-0.78539816d0-0.04166397d0*y
& -0.00003954d0*(y**2d0)+0.00262573d0*(y**3d0)
& -0.00054125d0*(y**4d0)-0.00029333d0*(y**5d0)
& +0.00013558d0*(y**6d0)
```

---

[3]Based on the polynomial formulae in *Handbook of Mathematical Functions* by M Abramowitz and IA Stegun (Dover, 1970). The magnitude of error in the each of the formulae is smaller than 10^{-7}.

```
      BY0=f0*dsin(th0)/dsqrt(x)
      endif

20    continue

      return
      end

      function BY1(x)

      double precision x,y,BY1,BJ1,f0,th0

      if (x.le.0) then
      print*,'Invalid argument in calculation
     & of Y1(x)'
      goto 30
      endif

      if (x.lt.3d0) then
      y=(x/3d0)**2d0
      BJ1=x*(0.5d0-0.56249985d0*y
     & +0.21093573d0*(y**2d0)
     & -0.03954289d0*(y**3d0)+0.00443319d0*(y**4d0)
     & -0.00031761d0*(y**5d0)+0.00001109d0*(y**6d0))
      BY1=(0.63661977d0*x*dlog(0.5d0*x)*BJ1
     & -0.6366198d0
     & +0.2212091d0*y+2.1682709d0*(y**2d0)
     & -1.3164827d0*(y**3d0)+0.3123951d0*(y**4d0)
     & -0.0400976d0*(y**5d0)+0.0027873d0*(y**6d0))/x
      else
      y=3d0/x
      f0=0.79788456d0+0.00000156d0*y
     & +0.01659667d0*(y**2d0)
     & +0.00017105d0*(y**3d0)-0.00249511d0*(y**4d0)
     & +0.00113653d0*(y**5d0)-0.00020033d0*(y**6d0)
      th0=x-2.35619449d0+0.12499612d0*y
     & +0.00005650d0*(y**2d0)-0.00637879d0*(y**3d0)
```

```fortran
    & +0.00074348d0*(y**4d0)+0.00079824d0*(y**5d0)
    & -0.00029166d0*(y**6d0)
      BY1=f0*dsin(th0)/dsqrt(x)
      endif


30 continue

   return
   end
```

The subroutine CPG listed below accepts w, ξ, η, $x^{(k)}$, $y^{(k)}$, $n_x^{(k)}$, $n_y^{(k)}$ and $\ell^{(k)}$ (w, xi, eta, xk, yk, nkx, nky and L) as input parameters and return the values of $\mathcal{G}_1^{(k)}(\xi,\eta)$ and $\mathcal{G}_2^{(k)}(\xi,\eta)$ (in G1 and G2). In CPG, we use $n=8$ in Eq. (3.23) to compute the integrals defining $\mathcal{G}_1^{(k)}(\xi,\eta)$ and $\mathcal{G}_2^{(k)}(\xi,\eta)$. Note that for $n=8$, $w_{i+4}=w_i$ and $x_{i+4}=-x_i$ for $i=1$, 2, 3, 4. The weights w_1, w_2, w_3 and w_4 and the abscissas x_1, x_2, x_3 and x_4 are stored in the arrays gw(1:4) and gx(1:4).

```fortran
   subroutine CPG(w,xi,eta,xk,yk,nkx,nky,L,G1,G2)

   double precision xi,eta,xk,yk,nkx,nky,L,G1,G2,
   & A,B,E,w,gw(4),gx(4),BY0,BY1,yip,yim

   data gx(1),gx(2),gx(3),gx(4)/
   & 0.183434642495650d0,
   & 0.525532409916329d0,0.796666477413627d0,
   & 0.960289856497536d0/
   data gw(1),gw(2),gw(3),gw(4)/
   & 0.362683783378362d0,
   & 0.313706645877887d0,0.222381034453374d0,
   & 0.101228536290376d0/

   A=L**2d0
   B=2d0*L*(-nky*(xk-xi)+nkx*(yk-eta))
   E=(xk-xi)**2d0+(yk-eta)**2d0
```

```
      G1=0d0
      G2=0d0
      do 10 i=1,4
      yip=0.25d0*gx(i)+0.25d0
      yim=-0.25d0*gx(i)+0.25d0
      G1=G1+gw(i)*(BY0(w*dsqrt(A*(yip**2d0)+B*yip+E))
     & +BY0(w*dsqrt(A*(yim**2d0)+B*yim+E)))
      yip=0.25d0*gx(i)+0.75d0
      yim=-0.25d0*gx(i)+0.75d0
      G1=G1+gw(i)*(BY0(w*dsqrt(A*(yip**2d0)+B*yip+E))
     & +BY0(w*dsqrt(A*(yim**2d0)+B*yim+E)))
10    continue
      G1=0.25d0*0.25d0*L*G1

      if (dabs(4d0*A*E-B**2d0).lt.0.000000000001d0)
     & goto 20

      do 15 i=1,4
      yip=0.25d0*gx(i)+0.25d0
      yim=-0.25d0*gx(i)+0.25d0
      G2=G2+gw(i)*(BY1(w*dsqrt(A*(yip**2d0)+B*yip+E))
     & /dsqrt(A*(yip**2d0)+B*yip+E)
     & +BY1(w*dsqrt(A*(yim**2d0)+B*yim+E))
     & /dsqrt(A*(yim**2d0)+B*yim+E))
      yip=0.25d0*gx(i)+0.75d0
      yim=-0.25d0*gx(i)+0.75d0
      G2=G2+gw(i)*(BY1(w*dsqrt(A*(yip**2d0)+B*yip+E))
     & /dsqrt(A*(yip**2d0)+B*yip+E)
     & +BY1(w*dsqrt(A*(yim**2d0)+B*yim+E))
     & /dsqrt(A*(yim**2d0)+B*yim+E))
15    continue
      G2=-0.25d0*L*0.25d0*(nkx*(xk-xi)
     & +nky*(yk-eta))*w*G2

20    continue

      return
```

```
end
```

The subroutine CEHHZ1 is a simple modification of CELAP1 to determine the values of ϕ and $\partial\phi/\partial n$ on all the boundary elements by solving Eqs. (3.18) and (3.19). The values of $\overline{\phi}^{(k)}$ and $\overline{p}^{(k)}$ are returned in the arrays phi(1:N) and dphi(1:N) respectively. The input parameters for the subroutine are as follows. The constant w is the real variable w. The integer N is the number of boundary elements. The midpoints $(\overline{x}^{(m)}, \overline{y}^{(m)})$ of the elements and the points $(x^{(k)}, y^{(k)})$ on the boundary and the lengths of the elements are stored in the real arrays xm(1:N), ym(1:N), xb(1:N), yb(1:N) and lg(1:N). As explained in Chapter 1, BCT(1:N) is the integer array recording the types of boundary conditions on the boundary elements, and BCV(1:N) is the real array storing the value of the quantity (either ϕ or $\partial\phi/\partial n$) specified on each element. For example, if ϕ is specified with a value of 0.50 on the element $C^{(20)}$ then BCT(20) and BCV(20) are given the values 0 and 0.50 respectively. On the other hand, if $\partial\phi/\partial n$ is specified with a value of 0.50 on $C^{(20)}$ then BCT(20) and BCV(20) are assigned 1 and 0.50 respectively. Note that the subroutine SOLVER for solving a system of linear algebraic equations is called in CEHHZ1.

```
      subroutine CEHHZ1(w,N,xm,ym,xb,yb,nx,ny,lg,
     & BCT,BCV,phi,dphi)

      integer m,k,N,BCT(1000)

      double precision xm(1000),ym(1000),xb(1000),
     & yb(1000),nx(1000),ny(1000),lg(1000),BCV(1000),
     & B(1000),del,phi(1000),dphi(1000),G1,G2,
     & A(1000,1000),Z(1000),w

      do 10 m=1,N
      B(m)=0d0
      do 5 k=1,N
      call CPG(w,xm(m),ym(m),xb(k),yb(k),n
```

```
&  x(k),ny(k),lg(k),G1,G2)
   if (k.eq.m) then
   del=1d0
   else
   del=0d0
   endif

   if (BCT(k).eq.0) then
   A(m,k)=-G1
   B(m)=B(m)+BCV(k)*(-G2+0.5d0*del)
   else
   A(m,k)=G2-0.5d0*del
   B(m)=B(m)+BCV(k)*G1
   endif
 5 continue
10 continue

   call solver(A,B,N,1,Z)

   do 15 m=1,N
   if (BCT(m).eq.0) then
   phi(m)=BCV(m)
   dphi(m)=Z(m)
   else
   phi(m)=Z(m)
   dphi(m)=BCV(m)
   endif
15 continue

   return
   end
```

After CEHHZ1 is called, the returned output numbers stored in the arrays phi(1:N) and dphi(1:N) are used in CEHHZ2 to compute the solution ϕ at any desired point (ξ, η) in the solution domain. The values of ξ and η are stored in the real variables xi and eta respectively, and the returned value of

$\phi(\xi, \eta)$ in the real variable `pint`. Note that `CEHHZ2` is also a simple modification of `CELAP2` in Chapter 1.

```
    subroutine CEHHZ2(w,N,xi,eta,xb,yb,nx,ny,lg,
& phi,dphi,pint)

    integer N,i

    double precision xi,eta,xb(1000),yb(1000),
& nx(1000),ny(1000),lg(1000),phi(1000),
& dphi(1000),pint,sum,G1,G2,w

    sum=0d0

    do 10 i=1,N
    call CPG(w,xi,eta,xb(i),yb(i),nx(i),ny(i),
& lg(i),G1,G2)
    sum=sum+phi(i)*G2-dphi(i)*G1
10  continue

    pint=sum

    return
    end
```

Example 3.1

For a particular problem, let us take $w = \sqrt{2}\pi/4$ in Eq. (3.3), the solution domain to be $0 < x < 1$, $0 < y < 1$, and the boundary conditions as

$$\phi(0, y) = 0 \text{ for } 0 < y < 1,$$

$$\phi(1, y) = \frac{\sqrt{2}}{2} \cos(\frac{\pi y}{4}) \text{ for } 0 < y < 1,$$

$$\left.\frac{\partial \phi}{\partial n}\right|_{y=0} = 0 \text{ for } 0 < x < 1,$$

$$\left.\frac{\partial \phi}{\partial n}\right|_{y=1} = -\frac{\pi\sqrt{2}}{8}\sin(\frac{\pi x}{4}) \text{ for } 0 < x < 1,$$

It is easy to verify by direct substitution that the exact solution of this particular boundary value problem is

$$\phi(x,y) = \sin(\frac{\pi x}{4})\cos(\frac{\pi y}{4}).$$

The boundary of the square solution domain and the boundary conditions are set up in the main program **EX3PT1** (listed below) which is to be compiled together with the subroutines **CEHHZ1**, **CEHHZ2**, **SOLVER** (with its supporting subprograms) and **CPG** and the functions **BY0** and **BY1** to create an executable code.

In the program **EX3PT1**, each side of the square is discretized into an equal number of boundary elements of the same length. The value of the coefficient w in Eq. (3.3) is stored in the real variable **w**. The subroutine **CEHZZ1** is first called to solve for the unknowns on the boundary, followed by call of **CEHZZ2** to compute ϕ at selected interior points.

```
program EX3PT1

integer NO,BCT(1000),N,i,ians

double precision xb(1000),yb(1000),xm(1000),
& ym(1000),nx(1000),ny(1000),lg(1000),
& BCV(1000),phi(1000),dphi(1000),pint,dl,
& xi,eta,pi,w

print*,'Enter number of elements per side
& (<251):'
read*,NO
N=4*NO

pi=4d0*datan(1d0)
dl=1d0/dfloat(NO)
```

96

```fortran
      w=pi*dsqrt(2d0)*0.25d0

      do 10 i=1,N0
      xb(i)=dfloat(i-1)*dl
      yb(i)=0d0
      xb(N0+i)=1d0
      yb(N0+i)=xb(i)
      xb(2*N0+i)=1d0-xb(i)
      yb(2*N0+i)=1d0
      xb(3*N0+i)=0d0
      yb(3*N0+i)=1d0-xb(i)
10    continue

      xb(N+1)=xb(1)
      yb(N+1)=yb(1)

      do 20 i=1,N
      xm(i)=0.5d0*(xb(i)+xb(i+1))
      ym(i)=0.5d0*(yb(i)+yb(i+1))
      lg(i)=dsqrt((xb(i+1)-xb(i))**2d0
     & +(yb(i+1)-yb(i))**2d0)
      nx(i)=(yb(i+1)-yb(i))/lg(i)
      ny(i)=(xb(i)-xb(i+1))/lg(i)
20    continue

      do 30 i=1,N
      if (i.le.N0) then
      BCT(i)=1
      BCV(i)=0d0
      else if ((i.gt.N0).and.(i.le.(2*N0))) then
      BCT(i)=0
      BCV(i)=dcos(pi*ym(i)*0.25d0)*0.5d0*dsqrt(2d0)
      else if ((i.gt.(2*N0)).and.(i.le.(3*N0))) then
      BCT(i)=1
      BCV(i)=-pi*0.125d0*dsqrt(2d0)
     & *dsin(pi*xm(i)*0.25d0)
      else
```

```
      BCT(i)=0
      BCV(i)=0d0
      endif
30 continue

      call CEHHZ1(w,N,xm,ym,xb,yb,nx,ny,lg,
   & BCT,BCV,phi,dphi)

50 print*,'Enter coordinates xi and eta of
   & an interior point:'

      read*,xi,eta

      call CEHHZ2(w,N,xi,eta,xb,yb,nx,ny,lg,
   & phi,dphi,pint)

      write(*,60)pint,dsin(pi*xi*0.25d0)
   & *dcos(pi*eta*0.25d0)

60 format('Numerical and exact values are:',
   & F14.6,' and',F14.6,' respectively')

      print*,'To continue with another point
   & enter 1:'
      read*,ians

      if (ians.eq.1) goto 50

      end
```

We execute the program **EX3PT1** to compute numerically the solution ϕ at various selected points in the interior of the solution domain. In Table 3.1, the numerical solution calculated using $N = 40$ and $N = 160$ is compared with the exact solution at selected points. It is obvious that the numerical values converge to the exact ones as the number of elements used is increased from 40 to 160 (that is, as the length of each elements is decreased from 0.10 to 0.0250 units).

Table 3.1

(ξ, η)	40 elements	160 elements	Exact
$(0.10, 0.20)$	0.077126	0.077437	0.077493
$(0.10, 0.30)$	0.076071	0.076253	0.076291
$(0.10, 0.40)$	0.074458	0.074588	0.074619
$(0.50, 0.20)$	0.377914	0.377965	0.377972
$(0.50, 0.30)$	0.372056	0.372102	0.372110
$(0.50, 0.40)$	0.363908	0.363946	0.363954
$(0.90, 0.20)$	0.641654	0.641489	0.641452
$(0.90, 0.30)$	0.631592	0.631526	0.631504
$(0.90, 0.40)$	0.617709	0.617678	0.617662

3.3 Helmholtz Type Equation with Variable Coefficients

3.3.1 Integral Formulation

It may be difficult (if not impossible) to derive in exact form a fundamental solution of Eq. (3.1) for the general case in which the coefficients α and g are varying smoothly from point to point in space. Consequently, we may not be able to reduce Eq. (3.1) to a boundary integral equation which has the form of Eq. (3.4).

The fundamental solution of the two-dimensional Laplace's equation, that is,

$$\Phi(x, y; \xi, \eta) = \frac{1}{4\pi} \ln[(x - \xi)^2 + (y - \eta)^2]. \qquad (3.24)$$

may, however, be applied to derive an integral formulation for Eq. (3.1) as follows.

If the functions ϕ_1 and ϕ_2 satisfy the Poisson-like equations

$$\frac{\partial^2 \phi_1}{\partial x^2} + \frac{\partial^2 \phi_1}{\partial y^2} = \sigma_1,$$

$$\frac{\partial^2 \phi_2}{\partial x^2} + \frac{\partial^2 \phi_2}{\partial y^2} = \sigma_2, \qquad (3.25)$$

in the region R bounded by the curve C then it may be shown that

$$\int_C [\phi_2 \frac{\partial \phi_1}{\partial n} - \phi_1 \frac{\partial \phi_2}{\partial n}] ds(x, y)$$

$$= \iint_R [\phi_2 \sigma_1 - \phi_1 \sigma_2] dx dy. \qquad (3.26)$$

Note that in general σ_1 and σ_2 may represent any expression such that $\phi_2 \sigma_1 - \phi_1 \sigma_2$ is integrable over R. For example, we may have $\sigma_1 = x^2 + y^2$ and $\sigma_2 = \phi_2 + \partial\phi_2/\partial x$.

The reciprocal relation in Eq. (3.26) may be proven as follows. Multiply the first equation in Eq. (3.25) by ϕ_2 and the second equation by ϕ_1, take the difference of the two equations, integrate the resulting equation over the region R and apply the two-dimensional version of the divergence theorem (as in Section 1.3 of Chapter 1) to obtain Eq. (3.26).

The reciprocal relation in Eq. (3.26) may be used to derive an integral equation for the Helmholtz type equation given by Eq. (3.1). To do this, we take ϕ_1 to be the fundamental solution of the two-dimensional Laplace's equation, that is, $\phi_1 = \Phi(x, y; \xi, \eta)$, where Φ is as defined in Eq. (3.24), and ϕ_2 to be the required solution of Eq. (3.1) satisfying the boundary conditions given by Eq. (3.2), that is, $\phi_2 = \phi$. It follows that $\sigma_1 = 0$ and $\sigma_2 = g(x, y) - \alpha(x, y)\phi(x, y)$.

For a point (ξ, η) lying in the interior of R, the reciprocal relation holds for $\phi_1 = \Phi(x, y; \xi, \eta)$ and $\phi_2 = \phi$ if we replace the region R and the curve C in Eq. (3.26) by the shaded region in Figure 1.2 (on page 13) and its boundary $C \cup C_\varepsilon$ respectively, that is,

$$\int_C [\Phi \frac{\partial \phi}{\partial n} - \phi \frac{\partial \Phi}{\partial n}] ds(x,y) = -\int_{C_\varepsilon} [\Phi \frac{\partial \phi}{\partial n} - \phi \frac{\partial \Phi}{\partial n}] ds(x,y)$$

$$+ \iint_{R \backslash R_\varepsilon} [g - \alpha \phi] \Phi \, dx dy, \quad (3.27)$$

where $R \backslash R_\varepsilon$ denotes the (shaded) region between C and C_ε, that is, the domain R without the circular region R_ε that is bounded by C_ε.

If we let $\varepsilon \to 0^+$ in Eq. (3.27) and proceed as in Section 1.4 of Chapter 1 to evaluate the line integral over C_ε as $\varepsilon \to 0^+$, we obtain

$$\phi(\xi, \eta) = \iint_R \Phi(x, y; \xi, \eta)[g(x,y) - \alpha(x,y)\phi(x,y)] dx dy$$

$$+ \int_C [\phi(x,y) \frac{\partial}{\partial n} (\Phi(x,y;\xi,\eta))$$

$$- \Phi(x,y;\xi,\eta) \frac{\partial}{\partial n} (\phi(x,y))] ds(x,y) \text{ for } (\xi,\eta) \in R.$$

$$(3.28)$$

If the exercise above is repeated for (ξ, η) lying on a smooth part of C then C and C_ε in Eq. (3.27) have to be respectively replaced by D and D_ε shown in Figure 1.3 on page 17 and R_ε is the part of R inside the circle $(x - \xi)^2 + (y - \eta)^2 = \varepsilon^2$. We then obtain

$$\frac{1}{2}\phi(\xi, \eta) = \iint_R \Phi(x, y; \xi, \eta)[g(x,y) - \alpha(x,y)\phi(x,y)] dx dy$$

$$+ \int_C [\phi(x,y) \frac{\partial}{\partial n} (\Phi(x,y;\xi,\eta))$$

$$- \Phi(x,y;\xi,\eta) \frac{\partial}{\partial n} (\phi(x,y))] ds(x,y)$$

for (ξ, η) lying on a smooth part of C.

$$(3.29)$$

Note that the double integral in Eqs. (3.28) and (3.29) is improper since $\Phi(x, y; \xi, \eta)$ is not well defined at $(x, y) = (\xi, \eta)$. Strictly speaking, the integration should be over the region R without an infinitesimal part containing the point (ξ, η). Similarly, for $(\xi, \eta) \in C$, the line integral over C should be interpreted as over C without an infinitesimal portion containing (ξ, η).

Eqs. (3.28) and (3.29) may be summarized into a single equation as given by

$$
\begin{aligned}
\lambda(\xi, \eta)&\phi(\xi, \eta) \\
= \ & \iint\limits_{R} \Phi(x, y; \xi, \eta)[g(x, y) - \alpha(x, y)\phi(x, y)]dxdy \\
& + \int\limits_{C} [\phi(x, y)\frac{\partial}{\partial n}(\Phi(x, y; \xi, \eta)) \\
& -\Phi(x, y; \xi, \eta)\frac{\partial}{\partial n}(\phi(x, y))]ds(x, y) \\
& \quad \text{for } (\xi, \eta) \in R \cup C, \quad (3.30)
\end{aligned}
$$

with $\lambda(\xi, \eta)$ as defined in Eq. (3.5).

Eq. (3.30) does not only contain the usual line integral over the boundary C but also a double integral over the entire solution domain R. Furthermore, the unknown function ϕ appears in the integrand of the double integral. We may discretize R into tiny domain elements (as in finite element methods) to treat the double integral. Nevertheless, the discretization of R into domain elements may be avoided if we use the so called dual-reciprocity method[4] to approximately convert the double integral into a line integral over C.

---

[4]Such an approach was apparently first introduced by CA Brebbia and D Nardini in the paper "Dynamic analysis in solid mechanics by an alternative boundary element procedure" in *International Journal of Soil Dynamics and Earthquake Engineering* (Volume 2, 1983, pp. 228-233). Their work was recently republished in the journal *Engineering Analysis with Boundary Elements* (Volume 24, 2000, pp. 513-518).

3.3.2 Approximation of Domain Integral

A radial basis function ρ centered about the point (a,b) is a function of the form $\rho(x,y;a,b) = G(r(x,y;a,b))$, where $r(x,y;a,b) = \sqrt{(x-a)^2 + (y-b)^2}$, that is, $\rho(x,y;a,b)$ is dependent only on the distance of (x,y) from (a,b).

For our discussion here, we take ρ to be specifically given by[5]

$$\rho(x,y;a,b) = 1 + r^2(x,y;a,b) + r^3(x,y;a,b). \qquad (3.31)$$

We approximate $g(x,y) - \alpha(x,y)\phi(x,y)$ using radial basis functions of the form given in Eq. (3.31) centered about a set of M selected points in $R \cup C$. Specifically, if the selected points are given by $(a^{(1)}, b^{(1)})$, $(a^{(2)}, b^{(2)})$, $\cdots$, $(a^{(M-1)}, b^{(M-1)})$ and $(a^{(M)}, b^{(M)})$, we make the approximation

$$g(x,y) - \alpha(x,y)\phi(x,y) \simeq \sum_{m=1}^{M} \beta^{(m)} \rho(x,y;a^{(m)}, b^{(m)}), \qquad (3.32)$$

where $\beta^{(m)}$ are coefficients to be determined. For a good approximation, the selected points have to be well spaced out with some of them lying on the boundary C.

With Eq. (3.32), the double integral in Eq. (3.30) may be approximated by

$$\iint\limits_{R} \Phi(x,y;\xi,\eta)[g(x,y) - \alpha(x,y)\phi(x,y)]dxdy$$

$$\simeq \sum_{m=1}^{M} \beta^{(m)} \iint\limits_{R} \Phi(x,y;\xi,\eta)\rho(x,y;a^{(m)}, b^{(m)})dxdy.$$

$$(3.33)$$

---

[5]This specific radial basis function was suggested by Y Zhang and S Zhu in the paper "On the choice of interpolation functions used in the dual-reciprocity boundary-element method" in *Engineering Analysis with Boundary Elements* (Volume 13, 1994, pp. 387-396). In general, it may perhaps be better to take $\rho = 1 + pr^2 + qr^3$ with the constants p and q suitably selected.

Now if we can find a function $\chi(x, y; a, b)$ such that

$$\frac{\partial^2 \chi}{\partial x^2} + \frac{\partial^2 \chi}{\partial y^2} = \rho, \tag{3.34}$$

then, using the reciprocal relation in Eq. (3.26) with $\phi_1 = \Phi(x, y; \xi, \eta)$ and $\phi_2 = \chi(x, y; a, b)$ (hence $\sigma_1 = 0$ and $\sigma_2 = \rho(x, y; a, b)$), we obtain

$$\iint_R \Phi(x, y; \xi, \eta) \rho(x, y; a, b) dx dy = \Psi(\xi, \eta; a, b), \tag{3.35}$$

where

$$\begin{aligned}
\Psi(\xi, &\eta; a, b) \\
= \quad &\lambda(\xi, \eta) \chi(\xi, \eta; a, b) + \int_C [\Phi(x, y; \xi, \eta) \frac{\partial}{\partial n}(\chi(x, y; a, b)) \\
&\qquad - \chi(x, y; a, b) \frac{\partial}{\partial n}(\Phi(x, y; \xi, \eta)] ds(x, y).
\end{aligned} \tag{3.36}$$

Can we find a function $\chi(x, y; a, b)$ satisfying Eq. (3.34) with $\rho(x, y; a, b)$ given by Eq. (3.31)? If we assume that the function $\chi(x, y; a, b)$ is also a radial basis function centered about (a, b), that is, if $\chi(x, y; a, b) = q(r)$ where r is the distance between (x, y) and (a, b), Eq. (3.34) with Eq. (3.31) may be rewritten (in polar coordinates) as

$$\frac{1}{r}\frac{d}{dr}(r\frac{\partial q}{\partial r}) = 1 + r^2 + r^3$$

which may then be integrated twice (with the constants of integration set to zero) to give

$$q(r) = \frac{1}{4}r^2 + \frac{1}{16}r^4 + \frac{1}{25}r^5.$$

Thus, a possible function for $\chi(x, y; a, b)$ is given by

$$\chi(x, y; a, b) = \frac{1}{4}r^2(x, y; a, b) + \frac{1}{16}r^4(x, y; a, b) + \frac{1}{25}r^5(x, y; a, b). \tag{3.37}$$

104

With Eq. (3.37), the double integral in Eq. (3.35), that is, $\Psi(\xi, \eta; a, b)$ as defined in Eq. (3.35), may be computed by evaluating a line integral with a known integrand over C. As we shall see later, the line integral may be approximately calculated by discretizing C into straight line segments.

If we let (x, y) in Eq. (3.32) be given in turn by the selected points $(a^{(1)}, b^{(1)})$, $(a^{(2)}, b^{(2)})$, $\cdots$, $(a^{(M-1)}, b^{(M-1)})$ and $(a^{(M)}, b^{(M)})$, we generate a system of M equations given by

$$
g(a^{(j)}, b^{(j)}) - \alpha(a^{(j)}, b^{(j)})\phi(a^{(j)}, b^{(j)})
$$
$$
= \sum_{m=1}^{M} \beta^{(m)} \rho(a^{(j)}, b^{(j)}; a^{(m)}, b^{(m)}) \text{ for } j = 1, 2, \cdots, M,
$$

which can be inverted to give

$$
\beta^{(m)} = \sum_{j=1}^{M} \omega^{(mj)} [g(a^{(j)}, b^{(j)}) - \alpha(a^{(j)}, b^{(j)})\phi(a^{(j)}, b^{(j)})]
$$
$$
\text{for } m = 1, 2, \cdots, M, \tag{3.38}
$$

where the coefficients $\omega^{(mj)}$ are defined by

$$
\sum_{j=1}^{M} \omega^{(kj)} \rho(a^{(j)}, b^{(j)}; a^{(m)}, b^{(m)}) = \begin{cases} 1 & \text{if } k = m, \\ 0 & \text{if } k \neq m, \end{cases} \tag{3.39}
$$

that is, if $\rho(a^{(j)}, b^{(j)}; a^{(m)}, b^{(m)})$ is the element in the j-th row and m-th column of the square matrix $\mathbf{Q}$ then $\omega^{(kj)}$ is the element in the k-th row and j-th column of $\mathbf{Q}^{-1}$.

From Eqs. (3.33), (3.35) and (3.38), we may write

$$
\iint_{R} \Phi(x, y; \xi, \eta)[g(x, y) - \alpha(x, y)\phi(x, y)] dx dy
$$
$$
\simeq \sum_{j=1}^{M} [g(a^{(j)}, b^{(j)}) - \alpha(a^{(j)}, b^{(j)})\phi(a^{(j)}, b^{(j)})]
$$
$$
\times \sum_{m=1}^{M} \omega^{(mj)} \Psi(\xi, \eta; a^{(m)}, b^{(m)}). \tag{3.40}
$$

3.3.3 Dual-reciprocity Boundary Element Procedure

With Eq. (3.40), we may rewrite Eq. (3.30) approximately as

$$
\begin{aligned}
\lambda(\xi,\eta)&\phi(\xi,\eta) \\
&= \sum_{j=1}^{M}[g(a^{(j)},b^{(j)}) - \alpha(a^{(j)},b^{(j)})\phi(a^{(j)},b^{(j)})] \\
&\times \sum_{m=1}^{M}\omega^{(mj)}\Psi(\xi,\eta;a^{(m)},b^{(m)}) \\
&+ \int_{C}[\phi(x,y)\frac{\partial}{\partial n}(\Phi(x,y;\xi,\eta)) \\
&-\Phi(x,y;\xi,\eta)\frac{\partial}{\partial n}(\phi(x,y))]ds(x,y) \\
&\qquad\qquad \text{for } (\xi,\eta) \in R \cup C.
\end{aligned} \qquad (3.41)
$$

The reciprocal relation in Eq. (3.26) is applied twice in the derivation of Eq. (3.41). The first time is in obtaining Eq. (3.30) and the second time in approximately converting the double integral in the integral equation into a line integral over C (that is, in deriving the approximate formula in Eq. (3.40)). Because of this, if a numerical solution for the boundary value problem defined by Eqs. (3.1) and (3.2) is obtained by using Eq. (3.41), it is termed a dual-reciprocity boundary element solution.

As before, we discretize C into N straight line elements $C^{(1)}$, $C^{(2)}, \cdots, C^{(N-1)}$ and $C^{(N)}$ and approximate ϕ and $\partial\phi/\partial n$ as constants $\overline{\phi}^{(k)}$ and $\overline{p}^{(k)}$ respectively on the element $C^{(k)}$. We take the midpoints of the boundary elements to be the first N collocation points in the approximation in Eq. (3.40). (The integer M is taken to be greater than N.) Specifically, let $(a^{(k)},b^{(k)}) = (\overline{x}^{(k)},\overline{y}^{(k)})$ (the midpoint of $C^{(k)}$) for $k = 1, 2, \cdots, N$. The remaining $M - N$ collocation points are chosen to be well spaced out points in the interior of R denoted by $(\overline{x}^{(N+j)},\overline{y}^{(N+j)})$ for $j = 1, 2, \cdots, M - N$.

Letting (ξ,η) in Eq. (3.41) to be given (in turn) by $(\overline{x}^{(n)},\overline{y}^{(n)})$

106

for $n = 1, 2, \cdots, N+L$ (where $L = M - N$ is the number of collocation points in the interior of R), we may proceed to obtain the approximation

$$\lambda(\overline{x}^{(n)}, \overline{y}^{(n)})\overline{\phi}^{(n)}$$

$$= \sum_{j=1}^{N+L} \mu^{(nj)}[g(\overline{x}^{(j)}, \overline{y}^{(j)}) - \alpha(\overline{x}^{(j)}, \overline{y}^{(j)})\overline{\phi}^{(j)}]$$

$$+ \sum_{k=1}^{N} \{\overline{\phi}^{(k)}\mathcal{F}_2^{(k)}(\overline{x}^{(n)}, \overline{y}^{(n)}) - \overline{p}^{(k)}\mathcal{F}_1^{(k)}(\overline{x}^{(n)}, \overline{y}^{(n)})\}$$

$$\text{for } n = 1, 2, \cdots, N+L, \qquad (3.42)$$

where $\overline{\phi}^{(n)} = (\overline{x}^{(n)}, \overline{y}^{(n)})$ $(n = 1, 2, \cdots, N+L)$ and

$$\mu^{(nj)} = \sum_{m=1}^{N+L} \omega^{(mj)} \Psi(\overline{x}^{(n)}, \overline{y}^{(n)}; \overline{x}^{(m)}, \overline{y}^{(m)}),$$

$$\mathcal{F}_1^{(k)}(\xi, \eta) = \int_{C^{(k)}} \Phi(x, y; \xi, \eta) ds(x, y),$$

$$\mathcal{F}_2^{(k)}(\xi, \eta) = \int_{C^{(k)}} \frac{\partial}{\partial n}[\Phi(x, y; \xi, \eta)] ds(x, y). \qquad (3.43)$$

Analytical formulae for calculating $\mathcal{F}_1^{(k)}(\xi, \eta)$ and $\mathcal{F}_2^{(k)}(\xi, \eta)$ are given in Chapter 1. To calculate $\mu^{(nj)}$, the function Ψ in Eq. (3.36) may be computed approximately by using

$$\Psi(\xi, \eta; a, b)$$

$$= \lambda(\xi, \eta)\chi(\xi, \eta; a, b)$$

$$+ \sum_{k=1}^{N} [n_x^{(k)}\frac{\partial}{\partial x}(\chi(x, y; a, b))$$

$$+ n_y^{(k)}\frac{\partial}{\partial y}(\chi(x, y; a, b))]\Big|_{(x,y)=(\overline{x}^{(k)}, \overline{y}^{(k)})} \mathcal{F}_1^{(k)}(\xi, \eta)$$

$$- \sum_{k=1}^{N} \chi(\overline{x}^{(k)}, \overline{y}^{(k)}; a, b)\mathcal{F}_2^{(k)}(\xi, \eta) \qquad (3.44)$$

Eq. (3.42) constitutes a system of $N + L$ linear algebraic equations in $N + L$ unknowns. The unknowns are either $\overline{\phi}^{(k)}$ or $\overline{p}^{(k)}$ (not both) for $k = 1, 2, \cdots, N$ (depending on whether ϕ or $\partial\phi/\partial n$ is specified on a boundary element) and $\overline{\phi}^{(N+j)}$ for $j = 1, 2, \cdots, L$ (the values of ϕ at the L interior collocation points). We rewrite Eq. (3.42) as

$$\sum_{k=1}^{N+L} a^{(nk)} z^{(k)}$$

$$= -\sum_{j=1}^{N+L} \mu^{(nj)} g(\overline{x}^{(j)}, \overline{y}^{(j)}) + \sum_{k=1}^{N} b^{(nk)}$$

$$\text{for } n = 1, 2, \cdots, N + L. \tag{3.45}$$

For $k = 1, 2, \cdots, N$, the coefficients $a^{(nk)}$, $b^{(nk)}$ and $z^{(k)}$ are given by

$$a^{(nk)} = \begin{cases} -\mathcal{F}_1^{(k)}(\overline{x}^{(n)}, \overline{y}^{(n)}) \text{ if } \phi \text{ is specified over } C^{(k)}, \\ \mathcal{F}_2^{(k)}(\overline{x}^{(n)}, \overline{y}^{(n)}) - \frac{1}{2}\delta^{(nk)} \\ -\mu^{(nk)}\alpha(\overline{x}^{(k)}, \overline{y}^{(k)}) \\ \qquad \text{if } \partial\phi/\partial n \text{ is specified over } C^{(k)}, \end{cases}$$

$$b^{(nk)} = \begin{cases} \overline{\phi}^{(k)}(-\mathcal{F}_2^{(k)}(\overline{x}^{(n)}, \overline{y}^{(n)}) + \frac{1}{2}\delta^{(nk)} \\ +\mu^{(nk)}\alpha(\overline{x}^{(k)}, \overline{y}^{(k)})) \text{ if } \phi \text{ is specified over } C^{(k)}, \\ \overline{p}^{(k)}\mathcal{F}_1^{(k)}(\overline{x}^{(n)}, \overline{y}^{(n)}) \text{ if } \partial\phi/\partial n \text{ is specified over } C^{(k)}, \end{cases}$$

$$z^{(k)} = \begin{cases} \overline{p}^{(k)} & \text{if } \phi \text{ is specified over } C^{(k)}, \\ \overline{\phi}^{(k)} & \text{if } \partial\phi/\partial n \text{ is specified over } C^{(k)}, \end{cases} \tag{3.46}$$

while, for $j = 1, 2, \cdots, L$, the coefficients $a^{(n[N+j])}$ and $z^{(N+j)}$ are given by

$$a^{(n[N+j])} = -\delta^{(n[N+j])} - \mu^{(n[N+j])}\alpha(\overline{x}^{(N+j)}, \overline{y}^{(N+j)})$$
$$z^{(N+j)} = \overline{\phi}^{(N+j)}. \tag{3.47}$$

As before, for $n = 1, 2, \cdots, N+L$ and $m = 1, 2, \cdots, N+L$, we define

$$\delta^{(nm)} = \begin{cases} 0 & \text{if } n \neq m, \\ 1 & \text{if } n = m. \end{cases} \qquad (3.48)$$

3.3.4 Implementation on Computer

FORTRAN functions RHO and CHI for computing the radial basis functions $\rho(x, y; a, b)$ and $\chi(x, y; a, b)$ respectively are listed below.

```
function RHO(x,y,a,b)

double precision RHO,x,y,a,b,r

r=dsqrt((x-a)**2d0+(y-b)**2d0)
RHO=1d0+r**2d0+r**3d0

return
end

function CHI(x,y,a,b)

double precision CHI,x,y,a,b,r

r=dsqrt((x-a)**2d0+(y-b)**2d0)
CHI=0.25d0*(r**2d0)+0.0625d0*(r**4d0)
& +0.04d0*(r**5d0)

return
end
```

We are also required to compute the directional derivative $\partial\chi/\partial n = n_x\partial\chi/\partial x + n_y\partial\chi/\partial y$. The function DCHI for doing this is listed below.

```
function DCHI(x,y,a,b,nx,ny)
```

```
double precision DCHI,x,y,a,b,nx,ny,r

r=dsqrt((x-a)**2d0+(y-b)**2d0)
DCHI=(0.5d0+0.25d0*(r**2d0)+0.2d0*(r**3d0))
& *((x-a)*nx+(y-b)*ny)

return
end
```

The values of $\mu^{(nj)}$, $\mathcal{F}_1^{(k)}(\overline{x}^{(n)}, \overline{y}^{(n)})$ and $\mathcal{F}_2^{(k)}(\overline{x}^{(n)}, \overline{y}^{(n)})$ as defined in Eq. (3.43) are required in setting up the system of linear algebraic equations in Eq. (3.45). We create a subroutine called CFMU to do this. The input parameters for the subroutine are N (number of boundary elements), L (number of interior collocation points), $(\overline{x}^{(n)}, \overline{y}^{(n)})$ (all the collocation points which include the midpoints of the boundary elements), $(x^{(k)}, y^{(k)})$ (all the starting points of the elements), $[n_x^{(k)}, n_y^{(k)}]$ (all the unit normal vectors to the elements) and $\ell^{(k)}$ (all the lengths of the elements). These are stored in the integer variables N and L, and the real arrays xm(1:N+L), ym(1:N+L), xb(1:N), yb(1:N), nx(1:N), ny(1:N) and lg(1:N). The output parameters $\mathcal{F}_1^{(k)}(\overline{x}^{(n)}, \overline{y}^{(n)})$ and $\mathcal{F}_2^{(k)}(\overline{x}^{(n)}, \overline{y}^{(n)})$ are in the array fa(1:N,1:N+L,1:2). For a particular example, the value of $\mathcal{F}_1^{(10)}(\overline{x}^{(20)}, \overline{y}^{(20)})$ is given in the variable fa(10,20,1). Lastly, the values of $\mu^{(nj)}$ are returned in the array mu(1:N+L,1:N+L). Note that the subroutine CPF is called in CFMU for calculating $\mathcal{F}_1^{(k)}(\overline{x}^{(n)}, \overline{y}^{(n)})$ and $\mathcal{F}_2^{(k)}(\overline{x}^{(n)}, \overline{y}^{(n)})$. The subroutine SOLVER is used to invert a square matrix in computing $\omega^{(kj)}$ as explained below Eq. (3.39).

```
subroutine CFMU(N,L,xm,ym,xb,yb,nx,ny,lg,fa,mu)

integer N,L,k,NL,ll,nn,mm,j

double precision xb(1000),yb(1000),xm(1000),
& ym(1000),nx(1000),ny(1000),lg(1000),
```

```fortran
     & fa(1000,1000,2),mu(1000,1000),PF1,PF2,pi,
     & comega(1000,1000),cc(1000,1000),RHO,
     & omega(1000,1000),PSI(1000,1000),lam,CHI,DCHI

       pi=4d0*datan(1d0)

       NL=N+L

       do 10 k=1,N
       do 10 nn=1,NL
       call CPF(xm(nn),ym(nn),xb(k),yb(k),nx(k),ny(k),
     & lg(k),PF1,PF2)
       fa(k,nn,1)=PF1/pi
       fa(k,nn,2)=PF2/pi
10     continue

       do 20 k=1,NL
       comega(k,k)=1d0
       do 20 nn=1,NL
       if (k.ne.nn) comega(k,nn)=0d0
       cc(k,nn)=RHO(xm(nn),ym(nn),xm(k),ym(k))
20     continue

       do 30 j=1,NL
       if (j.eq.1) then
       l1=1
       else
       l1=0
       endif
       call SOLVER(cc,comega(1,j),NL,l1,omega(1,j))
30     continue

       do 40 nn=1,NL
       if (nn.le.(N)) then
       lam=0.5d0
       else
       lam=1d0
```

```
      endif
      do 40 mm=1,NL
      PSI(nn,mm)=lam*CHI(xm(nn),ym(nn),xm(mm),ym(mm))
      do 40 k=1,N
      PSI(nn,mm)=PSI(nn,mm)
    & -CHI(xm(k),ym(k),xm(mm),ym(mm))*fa(k,nn,2)
    & +DCHI(xm(k),ym(k),xm(mm),ym(mm),nx(k),ny(k))
    & *fa(k,nn,1)
   40 continue

      do 50 nn=1,NL
      do 50 j=1,NL
      mu(nn,j)=0d0
      do 50 mm=1,NL
      mu(nn,j)=mu(nn,j)+omega(mm,j)*PSI(nn,mm)
   50 continue

      return
      end
```

The subroutine CEDRHZT solves the boundary value problem defined by Eqs. (3.1) and (3.2). The input parameters for the subroutine are N (the number of boundary elements, stored in the integer variable N), L (the number of interior collocation points, in the integer variable L), $\alpha(\overline{x}^{(n)}, \overline{y}^{(n)})$ and $g(\overline{x}^{(n)}, \overline{y}^{(n)})$ (the values of the coefficient α and g at all collocation points, in the real arrays alpha(1:N+L,1:N+L) amd gc(1:N+L,1:N+L)), $(\overline{x}^{(n)}, \overline{y}^{(n)})$ (all the collocation points, in the real arrays xm(1:N+L) and ym(1:N+L)), $(x^{(k)}, y^{(k)})$ (all the starting points of the elements, in the real arrays xb(1:N) and yb(1:N)), $[n_x^{(k)}, n_y^{(k)}]$ (all the unit normal vectors to the elements, in the real arrays nx(1:N+L) and ny(1:N+L)) and $\ell^{(k)}$ (all the lengths of the elements, in the real array lg(1:N)), the types of boundary conditions on the elements (in the integer array BCT(1:N)) and the values of the specified quantities on the boundary elements (in the real array BCV(1:N)). (Refer to page 93.) The approximate values of the solution ϕ at all the collocation points are returned in the real array phi(1:N+L). For example, the values of $\phi(\overline{x}^{(1)}, \overline{y}^{(1)})$ and $\phi(\overline{x}^{(20)}, \overline{y}^{(20)})$ are stored in phi(1) and phi(20) respectively.

```
      subroutine CEDRHZT(N,L,alpha,gc,xm,ym,xb,yb,
     & nx,ny,lg,BCT,BCV,phi)

      integer N,L,k,NL,nn,mm,j,BCT(1000)

      double precision xb(1000),yb(1000),xm(1000),
     & ym(1000),gc(1000),nx(1000),ny(1000),lg(1000),
     & fa(1000,1000,2),BCV(1000),phi(1000),
     & A(1000,1000),alpha(1000),B(1000),d1,
     & Z(1000),mu(1000,1000)

      call CFMU(N,L,xm,ym,xb,yb,nx,ny,lg,fa,mu)

      NL=N+L

      do 60 nn=1,NL
      do 60 k=1,NL
      A(nn,k)=0d0
   60 continue

      do 80 nn=1,NL
      do 70 k=1,N
      if (BCT(k).eq.0) then
      A(nn,k)=-fa(k,nn,1)
      else
      if (k.eq.nn) then
      d1=0.5d0
      else
      d1=0d0
      endif
      A(nn,k)=fa(k,nn,2)-d1-mu(nn,k)*alpha(k)
      endif
   70 continue
```

```
      do 75 j=1,L
      if (nn.eq.(N+j)) then
      d1=1d0
      else
      d1=0d0
      endif
      A(nn,N+j)=-d1-mu(nn,N+j)*alpha(N+j)
75    continue
80    continue

      do 180 nn=1,NL
      B(nn)=0d0
      do 170 k=1,N
      if (BCT(k).eq.0) then
      if (k.eq.nn) then
      d1=0.5d0
      else
      d1=0d0
      endif
      B(nn)=B(nn)+BCV(k)*(-fa(k,nn,2)
   &  +d1+mu(nn,k)*alpha(k))
      else
      B(nn)=B(nn)+BCV(k)*fa(k,nn,1)
      endif
170   continue
      do 173 j=1,NL
      B(nn)=B(nn)-mu(nn,j)*gc(j)
173   continue
180   continue

      call solver(A,B,NL,1,Z)

      do 190 k=1,N
      if (BCT(k).eq.0) then
      phi(k)=BCV(k)
      else
      phi(k)=Z(k)
```

```
      endif
190 continue

    do 200 k=1,L
    phi(N+k)=Z(N+k)
200 continue

    return
    end
```

Example 3.2

As in Example 3.1, let us take the solution domain to be $0 < x < 1$, $0 < y < 1$. The coefficients α and g in Eq. (3.1) are respectively taken to be given by

$$\alpha(x,y) = \frac{\pi^2}{8} + y,$$

$$g(x,y) = \frac{\pi^2}{8}xy + y\sin(\frac{\pi x}{4})\cos(\frac{\pi y}{4}) + xy^2.$$

The boundary conditions are

$$\phi(0,y) = 0 \text{ for } 0 < y < 1,$$

$$\phi(1,y) = \frac{\sqrt{2}}{2}\cos(\frac{\pi y}{4}) + y \text{ for } 0 < y < 1,$$

$$\left.\frac{\partial\phi}{\partial n}\right|_{y=0} = -x \text{ for } 0 < x < 1,$$

$$\left.\frac{\partial\phi}{\partial n}\right|_{y=1} = -\frac{\pi\sqrt{2}}{8}\sin(\frac{\pi x}{4}) + x \text{ for } 0 < x < 1,$$

It is easy to verify by direct substitution that the exact solution to this particular boundary value problem is

$$\phi(x,y) = \sin(\frac{\pi x}{4})\cos(\frac{\pi y}{4}) + xy.$$

115

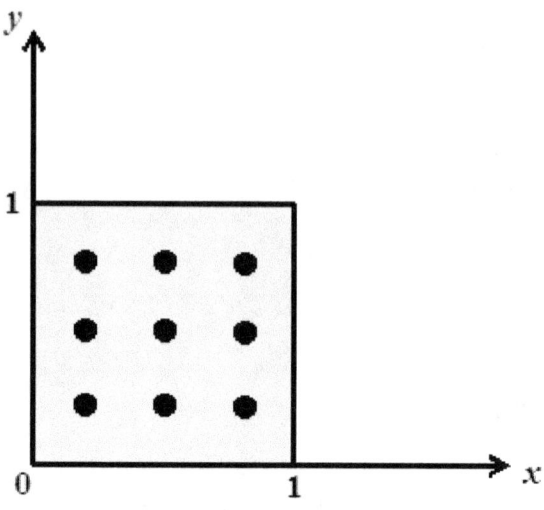

Figure 3.1

As in Example 3.1, each of the sides of the square is discretized into equal number of boundary elements of the same length. After discretizing the boundary into N elements, we choose the first N collocation points $(\overline{x}^{(1)}, \overline{y}^{(1)})$, $(\overline{x}^{(2)}, \overline{y}^{(2)})$, $\cdots$, $(\overline{x}^{(N-1)}, \overline{y}^{(N-1)})$ and $(\overline{x}^{(N)}, \overline{y}^{(N)})$ to be the midpoints of the elements. Another L collocation points $(\overline{x}^{(N+1)}, \overline{y}^{(N+1)})$, $(\overline{x}^{(N+2)}, \overline{y}^{(N+2)})$, $\cdots$, $(\overline{x}^{(N+L-1)}, \overline{y}^{(N+L-1)})$ and $(\overline{x}^{(N+L)}, \overline{y}^{(N+L)})$ are chosen to be well spaced out points in the interior of the square. More specifically, the interior collocation points are chosen to be given by $(k/(N_1 + 1), m/(N_1 + 1))$ for $k = 1, 2, \cdots, N_1$ and $m = 1, 2, \cdots, N_1$. Thus, there are N_1^2 evenly spaced out interior collocation points, that is, $L = N_1^2$. Figure 3.1 depicts the distribution of nine interior collocation points generated by using $N_1 = 3$.

The values of α and g at all the $N + L$ collocation points must be given as input data. In the program EX3PT2, these input numbers are stored in the real arrays alpha(1:N+L) and gc(1:N+L). After setting up all the required input data for the subroutine CEDRHZT, the program EX3PT2 calls the subroutine to obtain and print out the values of ϕ at all collocation points.

A complete listing of the program EX3PT2 is given below.

```fortran
      program EX3PT2

      integer N0,BCT(1000),N,i,ians,N1,L,NL,j,k

      double precision xb(1000),yb(1000),xm(1000)
     & ,ym(1000),dl,alpha(1000),nx(1000),ny(1000),
     & lg(1000),BCV(1000),pi,phi(1000),gc(1000)

      print*,'Enter integer N0 (no.  of elements
     & per side) (<200):'
      read*,N0
      N=4*N0

      print*,'Enter integer N1 (<15):'
      read*,N1
      L=N1**2
      NL=N+L

      pi=4d0*datan(1d0)

      dl=1d0/dfloat(N0)
      do 10 i=1,N0
      xb(i)=dfloat(i-1)*dl
      yb(i)=0d0
      xb(N0+i)=1d0
      yb(N0+i)=xb(i)
      xb(2*N0+i)=1d0-xb(i)
      yb(2*N0+i)=1d0
      xb(3*N0+i)=0d0
      yb(3*N0+i)=xb(2*N0+i)
 10   continue
      xb(N+1)=xb(1)
      yb(N+1)=yb(1)

      do 20 i=1,N
      xm(i)=0.5d0*(xb(i+1)+xb(i))
```

```fortran
      ym(i)=0.5d0*(yb(i+1)+yb(i))
      lg(i)=dsqrt((xb(i+1)-xb(i))**2d0
     & +(yb(i+1)-yb(i))**2d0)
      nx(i)=(yb(i+1)-yb(i))/lg(i)
      ny(i)=(xb(i)-xb(i+1))/lg(i)
20    continue

      dl=1d0/dfloat(N1+1)
      i=0
      do 25 j=1,N1
      do 25 k=1,N1
      i=i+1
      xm(N+i)=dfloat(j)*dl
      ym(N+i)=dfloat(k)*dl
25    continue

      do 26 i=1,N
      if ((i.le.N0).or.((i.gt.(2*N0))
     & .and.(i.le.(3*N0)))) then
      BCT(i)=1
      else
      BCT(i)=0
      endif
26    continue

      do 27 i=1,NL
      alpha(i)=pi*pi/8d0+ym(i)
      gc(i)=pi*pi*xm(i)*ym(i)/8d0
     & +xm(i)*ym(i)*ym(i)
     & +ym(i)*dsin(0.25d0*pi*xm(i))
     & *dcos(0.25d0*pi*ym(i))
27    continue

      do 30 i=1,N
      if (i.le.N0) then
      BCV(i)=0d0-xm(i)
      else if (i.le.(2*N0)) then
```

```
      BCV(i)=0.5d0*dsqrt(2d0)
   & *dcos(0.25d0*pi*ym(i))+ym(i)
      else if (i.le.(3*N0)) then
      BCV(i)=-pi*dsqrt(2d0)
   & *dsin(0.25d0*pi*xm(i))/8d0+xm(i)
      else
      BCV(i)=0d0
      endif
30 continue

      call CEDRHZT(N,L,alpha,gc,xm,ym,
   & xb,yb,nx,ny,lg,BCT,BCV,phi)

      print*,' x y Numerical Exact'
      print*,'------------------------'
      do 50 i=N+1,N+L
      write(*,60) xm(i),ym(i),phi(i),
   & dsin(0.25d0*pi*xm(i))
   & *dcos(0.25d0*pi*ym(i))+xm(i)*ym(i)
50 continue

60 format(4F14.6)

      end
```

The program EX3PT2 is compiled together with the functions RHO, CHI and DCHI and the subroutines CPF, CFMU, SOLVER (with its supporting subprograms) and CEDRHZT into an executable code.

In Table 3.2, we compare two sets of numerical values of ϕ with the exact solution at some collocation points in the interior of the solution domain. The first set is obtained by using 20 boundary elements and 9 interior collocation points, while the second one by 80 boundary elements and 49 interior collocation points. The numerical and the exact values are in good agreement with each other.

119

Table 3.2

(x, y)	$N = 20$ $L = 9$	$N = 80$ $L = 49$	Exact
$(0.25, 0.25)$	0.253056	0.253719	0.253842
$(0.25, 0.50)$	0.304853	0.305181	0.305240
$(0.25, 0.75)$	0.348407	0.349539	0.349712
$(0.50, 0.25)$	0.500857	0.500329	0.500330
$(0.50, 0.50)$	0.603669	0.603559	0.603557
$(0.50, 0.75)$	0.692971	0.693205	0.693190
$(0.75, 0.25)$	0.734409	0.732532	0.732395
$(0.75, 0.50)$	0.888856	0.888347	0.888280
$(0.75, 0.75)$	1.025066	1.023650	1.024440

3.4   Summary and Discussion

The derivation of a boundary element solution for the two-dimensional homogeneous Helmholtz equation is considered. Proceeding as in Chapter 1, we find that the fundamental solution of the Helmholtz equation may be given by a special function in the form of a zeroth order Bessel function of the second kind. With the fundamental solution, a boundary integral solution is obtained and discretized using constant elements. The integration of the fundamental solution and its normal derivative over each of the boundary elements is done numerically using the Gaussian integration formula. Apart from this, the boundary element procedure is essentially the same as the one described in Chapter 1.

It may not be possible to obtain a boundary integral solution for the more general Helmholtz type equation with variable coefficients. By using the fundamental solution of the Laplace's equation, an integral formulation may, however, be obtained. In addition to the usual boundary integral, the formulation contains a double integral over the entire solution domain. The double integral can be approximately converted to a line integral over the boundary of the solution domain by using the dual-reciprocity method. This gives rise to the dual-reciprocity boundary element method which does not require the solution domain to be discretized into domain elements. Only the boundary has to be discretized. Nevertheless, the unknowns in the resulting system of linear algebraic equations are not only restricted to quantities on the boundary, but are also the values of the unknown function at selected points in the interior of the solution domain.

The choice of the radial basis functions which may be used in the treatment of the double integral is not unique. The radial basis function $\rho(x, y; a, b) = 1 + r(x, y; a, b)$ was popular among researchers during the early stage in the development of the dual-reciprocity boundary element method. A drawback of this function is that it is not differentiable at the point $(x, y) = (a, b)$. It may give rise to significant errors in the numerical procedure for certain cases, such as when the line integral in Eq. (3.44) is computed by approximating χ and $\partial\chi/\partial n$ as constants over a boundary element. The radial basis function in Eq. (3.31) is used here because it is still relatively simple in form but has been proven capable of giving reasonably accurate results[6]. Many papers addressing both practical and theoretical issues on the use of radial basis functions in boundary element techniques can be found in the research literature[7].

The two-dimensional homogeneous Helmholtz equation may, of course, be solved by using the dual-reciprocity boundary element method too, if we wish to avoid dealing with the Bessel functions. The dual-reciprocity approach allows the boundary element method to be used for solving a wider class of problems[8], even when it is not possible to obtain analytically a fun-

---

[6] For an early account of this radial basis function, refer to the paper "On the choice of interpolation functions used in the dual-reciprocity boundary-element method" (by Zhang and S Zhu) in *Engineering Analysis with Boundary Elements* (Volume 13, 1994, pp. 387-396).

[7] See, for example, articles in *Engineering Analysis with Boundary Elements* (Volume 24, Numbers 7 and 8, 2000) (special issue on "The dual reciprocity method and radial basis functions").

[8] An example is the paper "A dual-reciprocity boundary element solution of a generalized non-linear Schrodinger equation" (by WT Ang and KC Ang) in *Numerical Methods for Partial Differential Equations* (Volume 20, 2004, pp. 843-854).

damental solution for the governing partial differential equation
of the problem under consideration.

3.5 Exercises

1. The two-dimensional wave equation is given by

$$\frac{\partial^2 u}{\partial x^2} + \frac{\partial^2 u}{\partial y^2} = \frac{1}{c^2}\frac{\partial^2 u}{\partial t^2} \quad (c \text{ is a positive real number})$$

where u is a function of the spatial variables x and y
and time t. Show that if $u(x,y,t) = \exp(i\omega t)\phi(x,y)$ ($i = \sqrt{-1}$) then the wave equation can be rewritten in the
form given by Eq. (3.3).

2. The two-dimensional steady-state heat conduction in a
functionally graded material is governed by the partial
differential equation

$$\frac{\partial}{\partial x}(k(x,y)\frac{\partial T}{\partial x}) + \frac{\partial}{\partial y}(k(x,y)\frac{\partial T}{\partial y}) = 0$$

where $T(x,y)$ is the temperature and $k(x,y)$ is the thermal conductivity. Assume that $k(x,y)$ is strictly positive and is partially differentiable twice with respect to
the spatial variables x and y in the domain of inter-
est. Show that if we make the substitution $T(x,y) = (k(x,y))^{-1/2}\phi(x,y)$ then the heat equation above can be
rewritten in the form given by Eq. (3.1).

3. Apply the Laplace transform (with respect to time t over
the interval $[0,\infty)$) on the wave equation in Exercise 1
to obtain a partial differential equation of the form given
by Eq. (3.1). (Hint. Multiply both sides of the wave
equation by $\exp(-pt)$ and integrate with respect to time
t over the interval $[0,\infty)$. Let

$$\phi(x,y;p) = \int_0^\infty u(x,y,t)\exp(-pt)dt.$$

Show that the resulting partial differential equation in ϕ assumes the form given in Eq. (3.1).)

4. [9]Find a suitable fundamental solution $\Omega(x, y; \xi, \eta)$ for the two-dimensional modified Helmholtz equation

$$\frac{\partial^2 \phi}{\partial x^2} + \frac{\partial^2 \phi}{\partial y^2} - w^2 \phi = 0 \quad (w \text{ is a positive real number})$$

such that a boundary integral solution in the form given by Eq. (3.4) can be obtained. Guided by Section 3.2, can you develop computer codes for a boundary element method which may be used to solve the modified Helmholtz equation subject to the boundary conditions in Eq. (3.2)?

5. Explain why $\rho(x, y; a, b) = 1 + r^2(x, y; a, b)$ may not be expected to work well for the approximation in Eq. (3.32).

6. Consider the boundary value problem which requires solving the partial differential equation

$$\frac{\partial^2 \phi}{\partial x^2} + \frac{\partial^2 \phi}{\partial y^2} + \phi = 0,$$

in the quarter-circular region $x^2 + y^2 < 1$, $x > 0$, $y > 0$, subject to the boundary conditions

$$\phi = 0 \text{ on } y = 0 \text{ for } 0 < x < 1,$$

$$\frac{\partial \phi}{\partial n} = 1 \text{ on } x^2 + y^2 = 1 \text{ for } x > 0 \text{ and } y > 0,$$

$$\phi = 0 \text{ on } x = 0 \text{ for } 0 < y < 1.$$

Use the subroutines CEHHZ1 and CEHHZ2 to compute ϕ numerically at the point $(x, y) = (\frac{1}{2} \cos \theta, \frac{1}{2} \sin \theta)$ for $\theta = \pi/6$, $\pi/4$ and $\pi/3$. Repeat the exercise using the subroutine CEDRHZT. Compare the numerical values of ϕ obtained using the two boundary element approaches at selected points in the interior of the quarter-circular region.

---

[9]The reference *Handbook of Mathematical Functions* by M Abramowitz and IA Stegun (Dover, 1970) may be useful in this exercise.

7. Use the subroutine **CEDRHZT** to solve numerically the ordinary differential equation in $\varphi(x)$ given by

$$\frac{d^2\varphi}{dx^2} + x\varphi = 2 + x + x^3 \quad \text{for } 0 < x < 1$$

subject to the conditions $\varphi(0) = 1$ and $\varphi(1) = 2$. (Hint. Recast the problem into one which requires solving the partial differential equation in the form of Eq. (3.1), that is, let $\phi(x,y) = \varphi(x)$. If the solution domain is taken to be the square region $0 < x < 1$, $0 < y < 1$, what conditions can you impose on the sides of the square? It is easy to verify that the exact solution is $\varphi(x) = 1 + x^2$. Check your numerical solution against the exact one.)

Chapter 4

Two-dimensional
Diffusion Equation

4.1   Introduction

The two-dimensional classical model for heat conduction or diffusion of substance gives rise to the parabolic partial differential equation[1]

$$\frac{\partial^2 \phi}{\partial x^2} + \frac{\partial^2 \phi}{\partial y^2} = \alpha \frac{\partial \phi}{\partial t}, \qquad (4.1)$$

where α is a given positive constant, t denotes time and $\phi(x, y, t)$ is temperature or concentration of matter.

For $t > 0$, we are interested in solving the diffusion (heat) equation in Eq. (4.1) in the two-dimensional region R bounded by a simple closed curve C (on the Oxy plane) subject to the

---

[1] For accounts on how the diffusion (heat) equation may be derived, the reader may refer to the following books: (a) *Conduction of Heat in Solids* by HS Carslaw and JC Jaeger (Oxford University Press, 1959), and (b) *The Mathematics of Diffusion* by J Crank (Oxford University Press, 1975). The more recent book *Industrial Mathematics: Case Studies in the Diffusion of Heat and Matter* by GR Fulford and P Broadbridge (Cambridge Press, 2002) contains some examples of industrial problems involving the equation.

initial-boundary conditions

$$\phi(x,y,0) = f_0(x,y) \text{ for } (x,y) \in R,$$
$$\phi(x,y,t) = f_1(x,y,t) \text{ for } (x,y) \in C_1,\ t > 0,$$
$$\frac{\partial}{\partial n}[\phi(x,y,t)] = f_2(x,y,t) \text{ for } (x,y) \in C_2,\ t > 0, \quad (4.2)$$

where f_0, f_1 and f_2 are suitably prescribed functions, C_1 and C_2 are non-intersecting curves such that $C_1 \cup C_2 = C$, $\partial\phi/\partial n = n_x \partial\phi/\partial x + n_y \partial\phi/\partial y$ and $[n_x, n_y]$ is the unit normal vector on C, pointing away from R. For a geometrical sketch of the problem, refer to Figure 1.1 on page 8.

The fundamental solution of the two-dimensional Laplace's equation as given (in Chapter 1) by

$$\Phi(x,y;\xi,\eta) = \frac{1}{4\pi}\ln[(x-\xi)^2 + (y-\eta)^2], \quad (4.3)$$

may be used to convert Eq. (4.1) to an integro-differential equation given by

$$\lambda(\xi,\eta)\phi(\xi,\eta,t)$$
$$= \iint\limits_{R} \alpha\Phi(x,y;\xi,\eta)\frac{\partial}{\partial t}[\phi(x,y,t)]dxdy$$
$$+ \int\limits_{C} [\phi(x,y,t)\frac{\partial}{\partial n}(\Phi(x,y;\xi,\eta))$$
$$-\Phi(x,y;\xi,\eta)\frac{\partial}{\partial n}(\phi(x,y,t))]ds(x,y)$$
$$\text{for } (\xi,\eta) \in R \cup C, \quad (4.4)$$

where

$$\lambda(\xi,\eta) = \begin{cases} 0 & \text{if } (\xi,\eta) \notin R \cup C, \\ 1/2 & \text{if } (\xi,\eta) \text{ lies on a smooth part of } C, \\ 1 & \text{if } (\xi,\eta) \in R, \end{cases} \quad (4.5)$$

Eq. (4.4) may be derived as explained in Section 3.3 (page 99, Chapter 3), with $g - \alpha\phi$ in Eqs. (3.27)-(3.30) replaced by $\alpha\partial\phi/\partial t$.

126

In this chapter, we show how Eq. (4.4) may be used to obtain a boundary element procedure for the numerical solution of the initial-boundary value problem defined by Eqs. (4.1) and (4.2). To avoid having to discretize the entire solution domain into many tiny cells, one may apply the dual-reciprocity method outlined in Section 3.3 to convert the domain integral in Eq. (4.4) approximately into a line integral over the boundary C. In implementing the boundary element procedure, only the boundary C has to be discretized into elements. Nevertheless, unlike in Chapters 1 and 2, the unknowns of the boundary element formulation here do not involve only the yet to be determined values of ϕ or $\partial\phi/\partial n$ on the boundary elements but also those of ϕ at selected collocation points in the interior of R.

The dual-reciprocity boundary element approach approximately reduces Eq. (4.4) into a system of linear equations containing unknown functions of time. First order time derivatives of some of the unknown functions are also present in the system. Several different approaches may be employed for solving the system of linear algebraic-differential equations. The approach used here is to approximate the first order time derivatives using a finite difference formula, so that the initial-boundary value problem can be formulated as systems of linear algebraic equations to be solved at consecutive time levels separated by a small time-step. For the approach to work well, the boundary element procedure used to obtain the numerical solution must be sufficiently accurate. For this reason, the discontinuous linear elements discussed in Chapter 2 will be employed for the approximation made on the boundary.

4.2 Solution by Dual-reciprocity Boundary Element Method

Following closely the analysis in Section 3.3, we can rewrite Eq. (4.4) approximately as

$$\lambda(\xi, \eta)\phi(\xi, \eta, t)$$

$$= \alpha \sum_{j=1}^{M} \frac{\partial}{\partial t}[\phi(x, y, t)]\Big|_{(x,y)=(a^{(j)}, b^{(j)})}$$

$$\times \sum_{m=1}^{M} \omega^{(mj)} \Psi(\xi, \eta; a^{(m)}, b^{(m)})$$

$$+ \int_C [\phi(x, y, t)\frac{\partial}{\partial n}(\Phi(x, y; \xi, \eta))$$

$$-\Phi(x, y; \xi, \eta)\frac{\partial}{\partial n}(\phi(x, y, t))]ds(x, y)$$

$$\text{for } (\xi, \eta) \in R \cup C. \qquad (4.6)$$

where $(a^{(1)}, b^{(1)})$, $(a^{(2)}, b^{(2)})$, $\cdots$, $(a^{(M-1)}, b^{(M-1)})$ and $(a^{(M)}, b^{(M)})$ are selected collocation points in $R \cup C$ and

$$\Psi(\xi, \eta; a, b)$$

$$= \lambda(\xi, \eta)\chi(\xi, \eta; a, b)$$

$$+ \int_C [\Phi(x, y; \xi, \eta)\frac{\partial}{\partial n}(\chi(x, y; a, b))$$

$$-\chi(x, y; a, b)\frac{\partial}{\partial n}(\Phi(x, y; \xi, \eta)]ds(x, y),$$

$$\sum_{j=1}^{M} \omega^{(kj)} \rho(a^{(j)}, b^{(j)}; a^{(m)}, b^{(m)}) = \begin{cases} 1 & \text{if } k = m, \\ 0 & \text{if } k \neq m, \end{cases}$$

$$\rho(x, y; a, b) = 1 + r^2(x, y; a, b) + r^3(x, y; a, b),$$

$$\chi(x, y; a, b) = \frac{1}{4}r^2(x, y; a, b) + \frac{1}{16}r^4(x, y; a, b) + \frac{1}{25}r^5(x, y; a, b),$$

$$r(x, y; a, b) = \sqrt{(x - a)^2 + (y - b)^2}. \qquad (4.7)$$

128

We approximate the boundary integral in Eq. (4.6) using the discontinuous linear elements as detailed in Chapter 2. To do this, we discretize C into N straight line elements $C^{(1)}$, $C^{(2)}$, $\cdots$, $C^{(N-1)}$ and $C^{(N)}$. The endpoints of the k-th element $C^{(k)}$ are denoted by $(x^{(k)}, y^{(k)})$ and $(x^{(k+1)}, y^{(k+1)})$. (Note that $(x^{(N+1)}, y^{(N+1)}) = (x^{(1)}, y^{(1)})$.) Two points $(\xi^{(k)}, \eta^{(k)})$ and $(\xi^{(N+k)}, \eta^{(N+k)})$ at a distance of $\tau \ell^{(k)}$ from $(x^{(k)}, y^{(k)})$ and $(x^{(k+1)}, y^{(k+1)})$ respectively, where τ is a positive number such that $0 < \tau < 1/2$ and $\ell^{(k)}$ is the length of $C^{(k)}$, are chosen on $C^{(k)}$. (Refer to Figure 2.1 on page 56.) For discontinuous linear elements, we make the approximation

$$
\phi(x, y, t)
$$
$$
\simeq \frac{[s(x,y) - (1-\tau)\ell^{(k)}]\widehat{\phi}^{(k)}(t) - [s(x,y) - \tau\ell^{(k)}]\widehat{\phi}^{(N+k)}(t)}{(2\tau - 1)\ell^{(k)}}
$$
$$
\text{for } (x,y) \in C^{(k)}, \tag{4.8}
$$

and

$$
\frac{\partial}{\partial n}[\phi(x, y, t)]
$$
$$
\simeq \frac{[s(x,y) - (1-\tau)\ell^{(k)}]\widehat{p}^{(k)}(t) - [s(x,y) - \tau\ell^{(k)}]\widehat{p}^{(N+k)}(t)}{(2\tau - 1)\ell^{(k)}}
$$
$$
\text{for } (x,y) \in C^{(k)}, \tag{4.9}
$$

where $\widehat{\phi}^{(k)}(t)$ and $\widehat{\phi}^{(N+k)}(t)$ are the values of $\phi(x,y,t)$ at $(x,y) = (\xi^{(k)}, \eta^{(k)})$ and $(x,y) = (\xi^{(N+k)}, \eta^{(N+k)})$ respectively, $\widehat{p}^{(k)}(t)$ and $\widehat{p}^{(N+k)}(t)$ are the values of $\partial[\phi(x,y,t)]/\partial n$ at $(x,y) = (\xi^{(k)}, \eta^{(k)})$ and $(x,y) = (\xi^{(N+k)}, \eta^{(N+k)})$ respectively, and

$$
s(x,y) = \sqrt{(x - x^{(k)})^2 + (y - y^{(k)})^2}
$$
$$
\text{for } (x,y) \in C^{(k)}.
$$

With Eqs. (4.8) and (4.9), we may write

$$
\int_C [\phi(x,y,t)\frac{\partial}{\partial n}(\Phi(x,y;\xi,\eta))
$$

$$
-\Phi(x,y;\xi,\eta)\frac{\partial}{\partial n}(\phi(x,y,t))]ds(x,y)
$$

$$
\simeq \sum_{k=1}^{N}\frac{1}{(2\tau-1)\ell^{(k)}}\Big\{\widehat{\phi}^{(k)}(t)
$$

$$
\times[-(1-\tau)\ell^{(k)}\mathcal{F}_2^{(k)}(\xi,\eta)+\mathcal{F}_4^{(k)}(\xi,\eta)]
$$

$$
+\widehat{\phi}^{(N+k)}(t)[\tau\ell^{(k)}\mathcal{F}_2^{(k)}(\xi,\eta)-\mathcal{F}_4^{(k)}(\xi,\eta)]
$$

$$
-\widehat{p}^{(k)}(t)[-(1-\tau)\ell^{(k)}\mathcal{F}_1^{(k)}(\xi,\eta)+\mathcal{F}_3^{(k)}(\xi,\eta)]
$$

$$
-\widehat{p}^{(N+k)}(t)[\tau\ell^{(k)}\mathcal{F}_1^{(k)}(\xi,\eta)-\mathcal{F}_3^{(k)}(\xi,\eta)]\Big\},
$$

where analytical formulae for calculating $\mathcal{F}_1^{(k)}(\xi,\eta)$, $\mathcal{F}_2^{(k)}(\xi,\eta)$, $\mathcal{F}_3^{(k)}(\xi,\eta)$ and $\mathcal{F}_4^{(k)}(\xi,\eta)$ are given in Section 2.3 of Chapter 2.

Note that $(a^{(1)},b^{(1)})$, $(a^{(2)},b^{(2)})$, $\cdots$, $(a^{(M-1)},b^{(M-1)})$ and $(a^{(M)},b^{(M)})$ in Eq. (4.6) are M selected points that are well spaced out in the region $R\cup C$. Taking $M=2N+L$, we choose the first $2N$ of these points to be those on the boundary elements given by $(\xi^{(k)},\eta^{(k)})$ and $(\xi^{(N+k)},\eta^{(N+k)})$ for $k=1$, 2, $\cdots$, N. The remaining L points $(\xi^{(2N+1)},\eta^{(2N+1)})$, $(\xi^{(2N+2)},\eta^{(2N+2)})$, $\cdots$, $(\xi^{(2N+L-1)},\eta^{(2N+L-1)})$ and $(\xi^{(2N+L)},\eta^{(2N+L)})$ are chosen points in the interior of R.

Below Eq. (4.9) we have defined

$$
\widehat{\phi}^{(n)}(t)=\phi(\xi^{(n)},\eta^{(n)},t)\ \text{for}\ n=1,2,\cdots,2N.
$$

In view of the L selected points inside R, we now extend the definition to include $n=2N+1, 2N+2, \cdots, 2N+L$.

Eq. (4.6) may now be approximately written as

$$
\begin{aligned}
& \lambda(\xi,\eta)\phi(\xi,\eta,t) \\
= \;& \alpha \sum_{j=1}^{2N+L} \frac{d}{dt}[\widehat{\phi}^{(j)}(t)] \sum_{m=1}^{2N+L} \omega^{(mj)} \Psi(\xi,\eta;\xi^{(m)},\eta^{(m)}) \\
& + \sum_{k=1}^{N} \frac{1}{(2\tau-1)\ell^{(k)}} \Big\{ \widehat{\phi}^{(k)}(t) \\
& \times [-(1-\tau)\ell^{(k)} \mathcal{F}_2^{(k)}(\xi,\eta) + \mathcal{F}_4^{(k)}(\xi,\eta)] \\
& + \widehat{\phi}^{(N+k)}(t)[\tau\ell^{(k)} \mathcal{F}_2^{(k)}(\xi,\eta) - \mathcal{F}_4^{(k)}(\xi,\eta)] \\
& - \widehat{p}^{(k)}(t)[-(1-\tau)\ell^{(k)} \mathcal{F}_1^{(k)}(\xi,\eta) + \mathcal{F}_3^{(k)}(\xi,\eta)] \\
& - \widehat{p}^{(N+k)}(t)[\tau\ell^{(k)} \mathcal{F}_1^{(k)}(\xi,\eta) - \mathcal{F}_3^{(k)}(\xi,\eta)] \Big\}. \quad (4.10)
\end{aligned}
$$

We assume that either ϕ or $\partial\phi/\partial n$ (not both) is specified across a given element. If ϕ is specified on $C^{(k)}$ then $\widehat{p}^{(k)}(t)$ and $\widehat{p}^{(N+k)}(t)$ are unknown functions. Otherwise, if $\partial\phi/\partial n$ is specified on $C^{(k)}$, $\widehat{\phi}^{(k)}(t)$ and $\widehat{\phi}^{(N+k)}(t)$ are unknowns. At the selected interior points $(\xi^{(2N+1)},\eta^{(2N+1)})$, $(\xi^{(2N+2)},\eta^{(2N+2)})$, $\cdots$, $(\xi^{(2N+L-1)},\eta^{(2N+L-1)})$ and $(\xi^{(2N+L)},\eta^{(2N+L)})$, ϕ is not known at all time except at $t=0$ [when it is given by the initial condition in Eq. (4.2)], that is, $\widehat{\phi}^{(2N+1)}(t)$, $\widehat{\phi}^{(2N+2)}(t)$, $\cdots$, $\widehat{\phi}^{(2N+L-1)}(t)$ and $\widehat{\phi}^{(2N+L)}(t)$ are unknown functions of t for $t>0$. Thus, there are $2N+L$ unknown functions of t on the right hand side of Eq. (4.10).

From Eq. (4.7), with the boundary C discretized into boundary elements, we may approximately evaluate $\Psi(\xi,\eta;a,b)$ using

$$
\begin{aligned}
& \Psi(\xi,\eta;a,b) \\
= \;& \lambda(\xi,\eta)\chi(\xi,\eta;a,b) \\
& - \sum_{k=1}^{N} \frac{1}{(2\tau-1)\ell^{(k)}} \Big\{ \chi(\xi^{(k)},\eta^{(k)};a,b) \\
& \times [-(1-\tau)\ell^{(k)} \mathcal{F}_2^{(k)}(\xi,\eta) + \mathcal{F}_4^{(k)}(\xi,\eta)] \\
& + \chi(\xi^{(N+k)},\eta^{(N+k)};a,b)[\tau\ell^{(k)} \mathcal{F}_2^{(k)}(\xi,\eta) - \mathcal{F}_4^{(k)}(\xi,\eta)]
\end{aligned}
$$

$$- \frac{\partial}{\partial n}[\chi(\xi, \eta; a, b)]\Big|_{(\xi,\eta)=(\xi^{(k)},\eta^{(k)})}$$
$$\times [-(1-\tau)\ell^{(k)}\mathcal{F}_1^{(k)}(\xi, \eta) + \mathcal{F}_3^{(k)}(\xi, \eta)]$$
$$- \frac{\partial}{\partial n}[\chi(\xi, \eta; a, b)]\Big|_{(\xi,\eta)=(\xi^{(N+k)},\eta^{(N+k)})}$$
$$\times [\tau\ell^{(k)}\mathcal{F}_1^{(k)}(\xi, \eta) - \mathcal{F}_3^{(k)}(\xi, \eta)]\Big\}. \qquad (4.11)$$

If we let (ξ, η) in Eq. (4.10) be given in turn by $(\xi^{(n)}, \eta^{(n)})$ for $n = 1, 2, \cdots, 2N + L$, we generate a system of $2N + L$ linear equations in $2N + L$ unknown functions of t, that is, we obtain

$$\lambda(\xi^{(n)}, \eta^{(n)})\widehat{\phi}^{(n)}(t)$$
$$= \alpha \sum_{j=1}^{2N+L} \mu^{(nj)}\frac{d}{dt}[\widehat{\phi}^{(j)}(t)]$$
$$+ \sum_{k=1}^{N} \frac{1}{(2\tau - 1)\ell^{(k)}} \Big\{\widehat{\phi}^{(k)}(t)$$
$$\times [-(1-\tau)\ell^{(k)}\mathcal{F}_2^{(k)}(\xi^{(n)}, \eta^{(n)}) + \mathcal{F}_4^{(k)}(\xi^{(n)}, \eta^{(n)})]$$
$$+ \widehat{\phi}^{(N+k)}(t)[\tau\ell^{(k)}\mathcal{F}_2^{(k)}(\xi^{(n)}, \eta^{(n)}) - \mathcal{F}_4^{(k)}(\xi^{(n)}, \eta^{(n)})]$$
$$- \widehat{p}^{(k)}(t)[-(1-\tau)\ell^{(k)}\mathcal{F}_1^{(k)}(\xi^{(n)}, \eta^{(n)}) + \mathcal{F}_3^{(k)}(\xi^{(n)}, \eta^{(n)})]$$
$$- \widehat{p}^{(N+k)}(t)[\tau\ell^{(k)}\mathcal{F}_1^{(k)}(\xi^{(n)}, \eta^{(n)}) - \mathcal{F}_3^{(k)}(\xi^{(n)}, \eta^{(n)})]\Big\}$$
$$\text{for } n = 1, 2, \cdots, 2N + L, \qquad (4.12)$$

where

$$\mu^{(nj)} = \sum_{m=1}^{2N+L} \omega^{(mj)}\Psi(\xi^{(n)}, \eta^{(n)}; \xi^{(m)}, \eta^{(m)}). \qquad (4.13)$$

Note that $\lambda(\xi^{(n)}, \eta^{(n)}) = 1/2$ for $n = 1, 2, \cdots, 2N$ and $\lambda(\xi^{(n)}, \eta^{(n)}) = 1$ for $n = 2N+1, 2N+2, \cdots, 2N+L$. The constants $\omega^{(mj)}$ are determined from Eq. (4.7) by letting $M = 2N + L$ and $(a^{(n)}, b^{(n)}) = (\xi^{(n)}, \eta^{(n)})$ for $n = 1, 2, \cdots, 2N+L$.

If we are able to solve Eq. (4.12) together with the initial-boundary conditions in Eq. (4.2), then we have determined ϕ numerically (for the initial-boundary value problem described in Section 3.1) at the $2N + L$ selected points in $R \cup C$.

4.3 Time-stepping Approach

In this section, we describe a time-stepping approach for solving Eq. (4.12) together with the initial-boundary conditions in Eq. (4.2).

The function $\widehat{\phi}^{(j)}(t)$ and its first order derivative are approximated using

$$
\begin{aligned}
\widehat{\phi}^{(j)}(t) &\simeq \frac{1}{2}[\widehat{\phi}^{(j)}(t + \frac{1}{2}\Delta t) + \widehat{\phi}^{(j)}(t - \frac{1}{2}\Delta t)], \\
\frac{d}{dt}[\widehat{\phi}^{(j)}(t)] &\simeq \frac{1}{\Delta t}[\widehat{\phi}^{(j)}(t + \frac{1}{2}\Delta t) - \widehat{\phi}^{(j)}(t - \frac{1}{2}\Delta t)],
\end{aligned}
\tag{4.14}
$$

where Δt is a small (positive) time-step. The errors in the approximations in Eq. (4.14) are of order $O([\Delta t]^2)$.

Substitution of Eqs. (4.14) into Eq. (4.12) yields

$$
\begin{aligned}
&\frac{1}{2}\lambda(\xi^{(n)}, \eta^{(n)})[\widehat{\phi}^{(n)}(t + \frac{1}{2}\Delta t) + \widehat{\phi}^{(n)}(t - \frac{1}{2}\Delta t)] \\
&= \frac{\alpha}{\Delta t} \sum_{j=1}^{2N+L} \mu^{(nj)}[\widehat{\phi}^{(j)}(t + \frac{1}{2}\Delta t) - \widehat{\phi}^{(j)}(t - \frac{1}{2}\Delta t)] \\
&+ \sum_{k=1}^{N} \frac{1}{(2\tau - 1)\ell^{(k)}} \left\{ \frac{1}{2}[\widehat{\phi}^{(k)}(t + \frac{1}{2}\Delta t) + \widehat{\phi}^{(k)}(t - \frac{1}{2}\Delta t)] \right. \\
&\times [-(1-\tau)\ell^{(k)}\mathcal{F}_2^{(k)}(\xi^{(n)}, \eta^{(n)}) + \mathcal{F}_4^{(k)}(\xi^{(n)}, \eta^{(n)})] \\
&+ \frac{1}{2}[\widehat{\phi}^{(N+k)}(t + \frac{1}{2}\Delta t) + \widehat{\phi}^{(N+k)}(t - \frac{1}{2}\Delta t)] \\
&\times [\tau\ell^{(k)}\mathcal{F}_2^{(k)}(\xi^{(n)}, \eta^{(n)}) - \mathcal{F}_4^{(k)}(\xi^{(n)}, \eta^{(n)})] \\
&- \widehat{p}^{(k)}(t)[-(1-\tau)\ell^{(k)}\mathcal{F}_1^{(k)}(\xi^{(n)}, \eta^{(n)}) + \mathcal{F}_3^{(k)}(\xi^{(n)}, \eta^{(n)})] \\
&\left. - \widehat{p}^{(N+k)}(t)\, [\tau\ell^{(k)}\mathcal{F}_1^{(k)}(\xi^{(n)}, \eta^{(n)}) - \mathcal{F}_3^{(k)}(\xi^{(n)}, \eta^{(n)})] \right\} \\
&\quad\text{for } n = 1, 2, \cdots, 2N + L,
\end{aligned}
\tag{4.15}
$$

If we assume that $\widehat{\phi}^{(j)}(t - \tfrac{1}{2}\Delta t)$ $(j = 1, 2, \cdots, 2N + L)$ are known then Eq. (4.15) constitutes a system of $2N + L$ linear algebraic equations containing $2N+L$ unknowns. There are $2N$ unknowns on the boundary. They are given by $\widehat{\phi}^{(k)}(t + \tfrac{1}{2}\Delta t)$ and $\widehat{\phi}^{(N+k)}(t + \tfrac{1}{2}\Delta t)$ if $\partial\phi/\partial n$ is specified on the boundary element $C^{(k)}$, or by $\widehat{p}^{(k)}(t)$ and $\widehat{p}^{(N+k)}(t)$ if ϕ is known on $C^{(k)}$. The remaining unknowns are the values of ϕ at the L chosen interior points, as given by $\widehat{\phi}^{(j)}(t+\tfrac{1}{2}\Delta t)$ for $j = 2N+1, 2N+2, \cdots, 2N + L$.

Eq. (4.15) may be solved at consecutive time levels $t = \tfrac{1}{2}\Delta t$, $\tfrac{3}{2}\Delta t$, $\tfrac{5}{2}\Delta t$, $\cdots$, as follows.

If we let $t = \tfrac{1}{2}\Delta t$, we find that $\widehat{\phi}^{(j)}(t - \tfrac{1}{2}\Delta t) = \widehat{\phi}^{(j)}(0)$ $(j = 1, 2, \cdots, 2N + L)$. For $j = 1, 2, \cdots, L$, it is obvious that $\widehat{\phi}^{(2N+j)}(0)$ can be determined directly from the initial condition in Eq. (4.2), that is, $\widehat{\phi}^{(2N+j)}(0) = f_0(\xi^{(2N+j)}, \eta^{(2N+j)})$. If there is no discontinuity between the value of ϕ in R at $t = 0$ and that specified on C_1 for $t > 0$, we may extend the initial condition to include all the points on the boundary and take $\widehat{\phi}^{(m)}(0) = f_0(\xi^{(m)}, \eta^{(m)})$ for $m = 1, 2, \cdots, 2N$. If a discontinuity exists, and if ϕ is known on $C^{(k)}$, then the known ϕ on $C^{(k)}$ for $t > 0$ is extended to include $t = 0$, so that $\widehat{\phi}^{(k)}(0)$ and $\widehat{\phi}^{(N+k)}(0)$ are respectively given by $f_1(\xi^{(k)}, \eta^{(k)}, 0)$ and $f_1(\xi^{(N+k)}, \eta^{(N+k)}, 0)$ [instead of $f_0(\xi^{(k)}, \eta^{(k)})$ and $f_0(\xi^{(N+k)}, \eta^{(N+k)})$], in order to ensure that $d[\widehat{\phi}^{(k)}(t)]/dt$ and $d[\widehat{\phi}^{(N+k)}(t)]/dt$ are well defined at $t = 0^+$. With $\widehat{\phi}^{(n)}(0)$ known for $n = 1, 2, \cdots, 2N + L$, we can solve Eq. (4.15) for the unknowns given by either $\widehat{\phi}^{(m)}(\Delta t)$ or $\widehat{p}^{(m)}(\tfrac{1}{2}\Delta t)$ for $m = 1, 2, \cdots, 2N$, and by $\widehat{\phi}^{(2N+j)}(\Delta t)$ for $j = 1, 2, \cdots, L$. Once these unknowns are determined, we can go on to the next time level $t = \tfrac{3}{2}\Delta t$ and solve Eq. (4.15) again for either $\widehat{\phi}^{(m)}(2\Delta t)$ or $\widehat{p}^{(m)}(\tfrac{3}{2}\Delta t)$ for $m = 1, 2, \cdots, 2N$, and $\widehat{\phi}^{(2N+j)}(2\Delta t)$ for $j = 1, 2, \cdots, L$. We may repeat the procedure at $t = \tfrac{5}{2}\Delta t$, $\tfrac{7}{2}\Delta t$, $\cdots$ to find the unknowns at higher time levels.

We may rewrite Eq. (4.15) as

$$\sum_{k=1}^{N}(a^{(nk)}z^{(k)}(t) + a^{(n[N+k])}z^{(N+k)}(t))$$

$$+ \sum_{j=1}^{L}a^{(n[2N+j])}z^{(2N+j)}(t)$$

$$= \sum_{k=1}^{N}[b^{(nk)}(t) + c^{(nk)}(t)]$$

$$+ \sum_{j=1}^{2N+L}[\frac{\alpha}{\Delta t}\mu^{(nj)} + \frac{1}{2}\delta^{(nj)}\lambda(\xi^{(j)}, \eta^{(j)})]\widehat{\phi}^{(j)}(t - \frac{1}{2}\Delta t)$$

$$\text{for } n = 1, 2, \cdots, 2N+L, \qquad (4.16)$$

where

$$a^{(nk)} = \begin{cases} (2\tau - 1)^{-1}[(1 - \tau)\mathcal{F}_1^{(k)}(\xi^{(n)}, \eta^{(n)}) \\ \quad -[\ell^{(k)}]^{-1}\mathcal{F}_3^{(k)}(\xi^{(n)}, \eta^{(n)})] \text{ if } \phi \text{ is given on } C^{(k)}, \\ (4\tau - 2)^{-1}[-(1 - \tau)\mathcal{F}_2^{(k)}(\xi^{(n)}, \eta^{(n)}) \\ \quad +[\ell^{(k)}]^{-1}\mathcal{F}_4^{(k)}(\xi^{(n)}, \eta^{(n)})] \\ \quad -\frac{1}{4}\delta^{(nk)} + \alpha\mu^{(nk)}(\Delta t)^{-1} \text{ if } \partial\phi/\partial n \text{ is given on } C^{(k)}, \end{cases}$$

$$a^{(n[N+k])} = \begin{cases} (2\tau - 1)^{-1}[-\tau\mathcal{F}_1^{(k)}(\xi^{(n)}, \eta^{(n)}) \\ \quad +[\ell^{(k)}]^{-1}\mathcal{F}_3^{(k)}(\xi^{(n)}, \eta^{(n)})] \text{ if } \phi \text{ is given on } C^{(k)}, \\ (4\tau - 2)^{-1}[\tau\mathcal{F}_2^{(k)}(\xi^{(n)}, \eta^{(n)}) \\ \quad -[\ell^{(k)}]^{-1}\mathcal{F}_4^{(k)}(\xi^{(n)}, \eta^{(n)})] \\ \quad -\frac{1}{4}\delta^{([n-N]k)} + \alpha\mu^{(n[N+k])}(\Delta t)^{-1} \\ \qquad \text{if } \partial\phi/\partial n \text{ is given on } C^{(k)}, \end{cases}$$

$$a^{(n[2N+j])} = \frac{\alpha}{\Delta t}\mu^{(n[2N+j])} - \frac{1}{2}\delta^{[n-2N]j} \text{ for } j = 1, 2, \cdots, L,$$

$$b^{(nk)}(t) = \begin{cases} -\widehat{\phi}^{(k)}(t + \frac{1}{2}\Delta t)\{(4\tau - 2)^{-1} \\ \times[-(1-\tau)\mathcal{F}_2^{(k)}(\xi^{(n)}, \eta^{(n)}) \\ +[\ell^{(k)}]^{-1}\mathcal{F}_4^{(k)}(\xi^{(n)}, \eta^{(n)})] \\ -\frac{1}{4}\delta^{(nk)} + \alpha\mu^{(nk)}(\Delta t)^{-1}\} \\ -\widehat{\phi}^{(N+k)}(t + \frac{1}{2}\Delta t)\{(4\tau - 2)^{-1}[\tau\mathcal{F}_2^{(k)}(\xi^{(n)}, \eta^{(n)}) \\ -[\ell^{(k)}]^{-1}\mathcal{F}_4^{(k)}(\xi^{(n)}, \eta^{(n)})] \\ -\frac{1}{4}\delta^{([n-N]k)} + \alpha\mu^{(n[N+k])}(\Delta t)^{-1}\} \\ \qquad\qquad \text{if } \phi \text{ is given on } C^{(k)}, \\ -\widehat{p}^{(k)}(t)(2\tau - 1)^{-1} \\ \times[(1-\tau)\mathcal{F}_1^{(k)}(\xi^{(n)}, \eta^{(n)}) \\ -[\ell^{(k)}]^{-1}\mathcal{F}_3^{(k)}(\xi^{(n)}, \eta^{(n)})] \\ -\widehat{p}^{(N+k)}(t)(2\tau - 1)^{-1}[-\tau\mathcal{F}_1^{(k)}(\xi^{(n)}, \eta^{(n)}) \\ +[\ell^{(k)}]^{-1}\mathcal{F}_3^{(k)}(\xi^{(n)}, \eta^{(n)})] \\ \qquad\qquad \text{if } \partial\phi/\partial n \text{ is given on } C^{(k)}, \end{cases}$$

$$\begin{aligned} c^{(nk)}(t) = \ & -\widehat{\phi}^{(k)}\left(t - \frac{1}{2}\Delta t\right)\{(4\tau - 2)^{-1} \\ & \times[-(1-\tau)\mathcal{F}_2^{(k)}(\xi^{(n)}, \eta^{(n)}) \\ & +[\ell^{(k)}]^{-1}\mathcal{F}_4^{(k)}(\xi^{(n)}, \eta^{(n)})]\} \\ & -\widehat{\phi}^{(N+k)}\left(t - \frac{1}{2}\Delta t\right)\{(4\tau - 2)^{-1}[\tau\mathcal{F}_2^{(k)}(\xi^{(n)}, \eta^{(n)}) \\ & -[\ell^{(k)}]^{-1}\mathcal{F}_4^{(k)}(\xi^{(n)}, \eta^{(n)})]\} \end{aligned}$$

$$\delta^{(nm)} = \begin{cases} 0 & \text{if } n \neq m, \\ 1 & \text{if } n = m, \end{cases}$$

$$z^{(k)}(t) = \begin{cases} \widehat{p}^{(k)}(t) \text{ if } \phi \text{ is given on } C^{(k)}, \\ \widehat{\phi}^{(k)}(t + \frac{1}{2}\Delta t) \text{ if } \partial\phi/\partial n \text{ is given on } C^{(k)}, \end{cases}$$

$$z^{(N+k)}(t) = \begin{cases} \widehat{p}^{(N+k)}(t) \text{ if } \phi \text{ is given on } C^{(k)}, \\ \widehat{\phi}^{(N+k)}(t + \frac{1}{2}\Delta t) \text{ if } \partial\phi/\partial n \text{ is given on } C^{(k)}, \end{cases}$$

$$z^{(2N+j)}(t) = \widehat{\phi}^{(2N+j)}\left(t + \frac{1}{2}\Delta t\right) \text{ for } j = 1, 2, \cdots, L. \quad (4.17)$$

4.4   Implementation on Computer

The radial basis functions $\rho(x, y; a, b)$ and $\chi(x, y; a, b)$ are programmed in the functions RHO and CHI. The normal derivative of χ, that is, $\partial\chi/\partial n = n_x\partial\chi/\partial x + n_y\partial\chi/\partial y$, is coded in the function DCHI. These functions are listed in Section 3.3 (page 109, Chapter 3).

As in Chapter 2, the number of boundary elements is stored in the integer variable N, the parameter τ which defines the discontinuous linear elements in the real variable tau, the boundary points $(x^{(m)}, y^{(m)})$ $(m = 1, 2, \cdots, N+1)$ in the real arrays xb(1:N+1) and yb(1:N+1), the points $(\xi^{(m)}, \eta^{(m)})$ $(m = 1, 2, \cdots, 2N)$ on the boundary elements in the real arrays xm(1:2*N) and ym(1:2*N), the unit normal vectors $(n_x^{(m)}, n_y^{(m)})$ $(m = 1, 2, \cdots, N)$ in the real arrays nx(1:N) and ny(1:N), and the lengths of the boundary elements in lg(1:N). Also, the boundary conditions on the elements are recorded by the integer array BCT(1:N) and the real array BCV(1:2*N), as explained on page 65 in Chapter 2. The collocation points $(\xi^{(m)}, \eta^{(m)})$ $(m = 2N + 1, 2N + 2, \cdots, 2N + L)$ are kept in the real arrays xm(2*N+1:2*N+L) and ym(2*N+1:2*N+L), if the number of interior collocation points is given in the integer variable L.

To set up Eq. (4.16), we have to calculate $\mathcal{F}_i^{(k)}(\xi^{(n)}, \eta^{(n)})$ $(k = 1, 2, \cdots, 2N; n = 1, 2, \cdots, 2N+L; i = 1, 2, 3, 4)$ and $\mu^{(nj)}$ $(n, j = 1, 2, \cdots, 2N + L)$. This is done in the subroutine DFMU which accepts N, L, tau, xm(1:2*N), ym(1:2*N), xb(1:N+1), yb(1:N+1), nx(1:N), ny(1:N) and lg(1:N) as input parameters. The values of $\mathcal{F}_i^{(k)}(\xi^{(n)}, \eta^{(n)})$ and $\mu^{(nj)}$ are returned in the real array fa(1:2*N,1:2*N+L,1:4) and mu(1:2*N+L,1:2*N+L) respectively.

The subroutine DFMU is listed below. Note that the subroutines DPF (page 64 in Chapter 2) and SOLVER (page 30 in Chapter 1) are called in DFMU.

```
      subroutine DFMU(N,L,tau,xm,ym,
     & xb,yb,nx,ny,lg,fa,mu)
```

```fortran
      integer N,L,k,NL,l1,nn,mm,j

      double precision xb(1000),yb(1000),
     & xm(1000),ym(1000),nx(1000),ny(1000),
     & lg(1000),fa(1000,1000,4),mu(1000,1000),
     & PF1,PF2,PF3,PF4,pi,comega(1000,1000),
     & cc(1000,1000),RHO,omega(1000,1000),
     & PSI(1000,1000),lam,CHI,DCHI,tau

       pi=4d0*datan(1d0)

      NL=2*N+L

      do 10 k=1,N
      do 10 nn=1,NL
      call DPF(xm(nn),ym(nn),xb(k),yb(k),
     & nx(k),ny(k),lg(k),
     & PF1,PF2,PF3,PF4)
       fa(k,nn,1)=PF1/pi
       fa(k,nn,2)=PF2/pi
       fa(k,nn,3)=PF3/pi
       fa(k,nn,4)=PF4/pi
 10   continue

      do 20 k=1,NL
      comega(k,k)=1d0
      do 20 nn=1,NL
      if (k.ne.nn) comega(k,nn)=0d0
      cc(k,nn)=RHO(xm(nn),ym(nn),xm(k),ym(k))
 20   continue

      do 30 j=1,NL
      if (j.eq.1) then
      l1=1
      else
      l1=0
      endif
```

```
      call SOLVER(cc,comega(1,j),NL,l1,omega(1,j))
30 continue

   do 40 nn=1,NL
   if (nn.le.(2*N)) then
   lam=0.5d0
   else
   lam=1d0
   endif
   do 40 mm=1,NL
   PSI(nn,mm)=lam*CHI(xm(nn),ym(nn),xm(mm),ym(mm))
   do 40 k=1,N
   PSI(nn,mm)=PSI(nn,mm)
 & -(CHI(xm(k),ym(k),xm(mm),ym(mm)))
 & *(-(1d0-tau)*lg(k)*fa(k,nn,2)+fa(k,nn,4))
 & +CHI(xm(N+k),ym(N+k),xm(mm),ym(mm))
 & *(tau*lg(k)*fa(k,nn,2)-fa(k,nn,4))
 & -DCHI(xm(k),ym(k),xm(mm),ym(mm),nx(k),ny(k))
 & *(-(1d0-tau)*lg(k)*fa(k,nn,1)+fa(k,nn,3))
 & -DCHI(xm(N+k),ym(N+k),xm(mm),ym(mm),
 & nx(k),ny(k))
 & *(tau*lg(k)*fa(k,nn,1)-fa(k,nn,3)))
 & /(lg(k)*(2d0*tau-1d0))
40 continue

   do 50 nn=1,NL
   do 50 j=1,NL
   mu(nn,j)=0d0
   do 50 mm=1,NL
   mu(nn,j)=mu(nn,j)+omega(mm,j)*PSI(nn,mm)
50 continue

   return
   end
```

A subroutine called `DLEDIFF` is created to set up and solve Eq. (4.16). The values in `tau`, `xm(1:2*N+L)`, `ym(1:2*N+L)`,

139

xb(1:N+1), yb(1:N+1), nx(1:N), ny(1:N), lg(1:N), BCT(1:N) and BCV(1:2*N) are accepted as inputs. Additional inputs are the values in the integer variable lud, the real variable alpha (the coefficient α in the diffusion equation), the real variable dt (the time-step Δt) and the real array phi(1:2*N+L). The variable phi(k) stores the value of ϕ at the collocation point $(\xi^{(k)}, \eta^{(k)})$ at time $t - \frac{1}{2}\Delta t$. We assume that $\phi(x, y, t - \frac{1}{2}\Delta t)$ is known and are interested in finding $\phi(x, t + \frac{1}{2}\Delta t)$ at the all the selected interior collocation points. Thus, BCV(k) and BCV(k+2*N) contain the values of either ϕ at time $t + \frac{1}{2}\Delta t$ or $\partial\phi/\partial n$ at time t (depending on whether ϕ or $\partial\phi/\partial n$ is specified on $C^{(k)}$). Once the subroutine DLEDIFF is executed, the values of the array phi(1:2*N+L) are altered to store the required values of $\phi(x, y, t + \frac{1}{2}\Delta t)$ at the collocation points (including those on the boundary elements). For example, the numerical value of ϕ at the interior point $(\xi^{(2N+3)}, \eta^{(2N+3)})$ at time $t + \frac{1}{2}\Delta t$ is kept in phi(2*N+3) after the calculation in DLEDIFF is completed.

The subroutine DLEDIFF may be called several times to find ϕ at higher and higher time levels. If the time-step, the boundary elements and the collocation points do not change for a particular problem, the coefficients $a^{(km)}$ $(k, m = 1, 2, \cdots, 2N+L)$ remain the same and need not be computed again at a higher time level. The re-calculation of $a^{(km)}$ can be avoided by giving the integer variable lud the value 0. Otherwise, lud can be given any other value.

When DLEDIFF is called the first time with $t = \frac{1}{2}\Delta t$, the real array phi(1:2*N+L) contains the values of $\widehat{\phi}^{(k)}(0)$ $(k = 1, 2, \cdots, 2N + L)$ as input, and BCV(1:2*N) contains the specified boundary values of either $\phi(\Delta t)$ or $\partial\phi/\partial n$ at time $\frac{1}{2}\Delta t$. Immediately after the first call of DLEDIFF, the values of $\phi(\xi^{(k)}, \eta^{(k)}, \Delta t)$ $(k = 1, 2, \cdots, 2N + L)$ are returned as output which may be used as input for the second call. The boundary values of either $\phi(2\Delta t)$ or $\partial\phi/\partial n$ at time $\frac{3}{2}\Delta t$ have to be stored in BCV(1:2*N) before DLEDIFF is called the second time with $t = \frac{3}{2}\Delta t$. The second call of DLEDIFF returns the values of $\phi(\xi^{(k)}, \eta^{(k)}, 2\Delta t)$ $(k = 1, 2, \cdots, 2N + L)$. In a similar manner,

DLEDIFF may be called again and again to obtain the values of ϕ at higher and higher time levels.

The subroutines DFMU (just created above) and SOLVER (page 30 in Chapter 1) and the functions RHO, CHI and DCHI are called in DLEDIFF listed below.

```
      subroutine DLEDIFF(lud,N,L,tau,alpha,xm,ym,
     & xb,yb,nx,ny,lg,BCT,BCV,dt,phi)

      integer N,L,k,NL,nn,mm,j,BCT(1000),lud

      double precision xb(1000),yb(1000),xm(1000),
     & ym(1000),nx(1000),ny(1000),lg(1000),tau,,
     & fa(1000,1000,4)BCV(1000),dt,phi(1000),
     & A(1000,1000),alpha,B(1000),d1,d2,Z(1000),
     & mu(1000,1000)

      common /AFMUPN/A,fa,mu,NL

      if (lud.eq.0) goto 90

      call DFMU(N,L,tau,xm,ym,xb,yb,nx,ny,lg,fa,mu)

      NL=2*N+L

      do 60 nn=1,NL
      do 60 k=1,NL
      A(nn,k)=0d0
60    continue

      do 80 nn=1,NL
      do 70 k=1,N
      if (BCT(k).eq.0) then
      A(nn,k)=((2d0*tau-1d0)**(-1d0))
     & *((1d0-tau)*fa(k,nn,1)
     & -(1d0/lg(k))*fa(k,nn,3))
      A(nn,k+N)=((2d0*tau-1d0)**(-1d0))
```

141

```fortran
     & *(-tau*fa(k,nn,1)
     & +(1d0/lg(k))*fa(k,nn,3))
       else
       if (k.eq.nn) then
       d1=1d0
       else
       d1=0d0
       endif
       if ((nn-N).eq.k) then
       d2=1d0
       else
       d2=0d0
       endif
       A(nn,k)=((4d0*tau-2d0)**(-1d0))
     & *(-(1d0-tau)*fa(k,nn,2)
     & +(1d0/lg(k))*fa(k,nn,4))
     & -0.25d0*d1+alpha*mu(nn,k)/dt
       A(nn,k+N)=((4d0*tau-2d0)**(-1d0))
     & *(tau*fa(k,nn,2)
     & -(1d0/lg(k))*fa(k,nn,4))
     & -0.25d0*d2+alpha*mu(nn,k+N)/dt
       endif
70   continue
       do 75 j=1,L
       if ((nn-2*N).eq.j) then
       d1=1d0
       else
       d1=0d0
       endif
       A(nn,2*N+j)=alpha*mu(nn,2*N+j)/dt-0.5d0*d1
75   continue
80   continue

90   continue

       do 180 nn=1,NL
       B(nn)=0d0
```

```fortran
      do 170 k=1,N
      if (BCT(k).eq.0) then
      if (k.eq.nn) then
      d1=1d0
      else
      d1=0d0
      endif
      if ((nn-N).eq.k) then
      d2=1d0
      else
      d2=0d0
      endif
      B(nn)=B(nn)-BCV(k)*( ((4d0*tau-2d0)**(-1d0))
     & *(-(1d0-tau)*fa(k,nn,2)
     & +(1d0/lg(k))*fa(k,nn,4))
     & -0.25d0*d1+alpha*mu(nn,k)/dt )
     & -BCV(k+N)*( ((4d0*tau-2d0)**(-1d0))
     & *(tau*fa(k,nn,2)-(1d0/lg(k))*fa(k,nn,4))
     & -0.25d0*d2+alpha*mu(nn,k+N)/dt )
      else
      B(nn)=B(nn)-BCV(k)*((2d0*tau-1d0)**(-1d0))
     & *((1d0-tau)*fa(k,nn,1)
     & -(1d0/lg(k))*fa(k,nn,3))
     & -BCV(k+N)*((2d0*tau-1d0)**(-1d0))
     & *(-tau*fa(k,nn,1)+(1d0/lg(k))*fa(k,nn,3))
      endif
      B(nn)=B(nn)-phi(k)*( ((4d0*tau-2d0)**(-1d0))
     & *(-(1d0-tau)*fa(k,nn,2)
     & +(1d0/lg(k))*fa(k,nn,4)))
     & -phi(k+N)*( ((4d0*tau-2d0)**(-1d0))
     & *(tau*fa(k,nn,2)-(1d0/lg(k))*fa(k,nn,4)))
170   continue
      do 173 j=1,NL
      if (j.eq.nn) then
      d1=1d0
      else
      d1=0d0
```

```
      endif
      if (j.le.(2*N)) then
      d2=0.5d0
      else
      d2=1d0
      endif
      B(nn)=B(nn)+(alpha*mu(nn,j)/dt
     & +0.5d0*d1*d2)*phi(j)
173   continue
180   continue

      call solver(A,B,NL,lud,Z)

      do 190 k=1,N
      if (BCT(k).eq.0) then
      phi(k)=BCV(k)
      phi(k+N)=BCV(k+N)
      else
      phi(k)=Z(k)
      phi(k+N)=Z(k+N)
      endif
190   continue

      do 200 k=1,L
      phi(2*N+k)=Z(2*N+k)
200   continue

      return
      end
```

4.5   Numerical Examples

Example 4.1

In the square region $0 < x < 1$, $0 < y < 1$, solve the diffusion equation

$$\frac{\partial^2 \phi}{\partial x^2} + \frac{\partial^2 \phi}{\partial y^2} = \frac{\partial \phi}{\partial t},$$

subject to the initial-boundary conditions

$$\phi = 1 + \cos(\frac{\pi x}{4})\sin(\frac{\pi y}{4}) \quad \text{at } t = 0$$
$$\text{for } 0 < x < 1,\, 0 < y < 1,$$

$$\phi = 1 + \exp(-\frac{\pi^2 t}{8})\sin(\frac{\pi y}{4}) \quad \text{on } x = 0$$
$$\text{for } 0 < y < 1,\, t > 0,$$

$$\phi = 1 + \frac{1}{\sqrt{2}}\exp(-\frac{\pi^2 t}{8})\sin(\frac{\pi y}{4}) \quad \text{on } x = 1$$
$$\text{for } 0 < y < 1,\, t > 0,$$

$$\frac{\partial \phi}{\partial n} = -\frac{\pi}{4}\exp(-\frac{\pi^2 t}{8})\cos(\frac{\pi x}{4}) \quad \text{on } y = 0$$
$$\text{for } 0 < x < 1,\, t > 0,$$

$$\frac{\partial \phi}{\partial n} = \frac{\pi}{4\sqrt{2}}\exp(-\frac{\pi^2 t}{8})\cos(\frac{\pi x}{4}) \quad \text{on } y = 1$$
$$\text{for } 0 < x < 1,\, t > 0.$$

The exact solution to the problem here is given by

$$\phi = 1 + \exp(-\frac{\pi^2 t}{8})\cos(\frac{\pi x}{4})\sin(\frac{\pi y}{4}).$$

To solve the problem numerically using DLEDIFF, each side of the square region is discretized into N_0 boundary elements of equal length (so that $N = 4N_0$). The interior collocation points for the dual-reciprocity method are chosen to be $(k/(N_1 + 1), m/(N_1 + 1))$ for $k = 1, 2, \cdots, N_1$ and $m = 1, 2, \cdots, N_1$. Thus, there are N_1^2 evenly spaced out interior collocation points, that is, $L = N_1^2$.

The main program EXP4PT1 for solving the test problem here is listed below. In EXP4PT1, the subroutine DLEDIFF may be called as many time as one likes. The first time when it is called at $t = \frac{1}{2}\Delta t$ the integer parameter `lud` is assigned the value 1. As the time-step Δt does not change and the same boundary elements and collocation points are used at higher time levels, `lud` is subsequently given the value 0, as the values in the real arrays `A(1:2*N+L)` do not have to be computed again. The numerical and exact values of ϕ at all the interior collocation points are printed out at each time level ($t = \Delta t$, $2\Delta t$, $3\Delta t$, $\cdots$).

```
program EX4PT1

integer N0,BCT(1000),N,i,ians,N1,L,NL,j,k,lud

double precision xb(1000),yb(1000),xm(1000),
& ym(1000),dl,alpha,nx(1000),ny(1000),lg(1000),
& BCV(1000),pi,tau,phi(1000),dt,ti,tir

print*,'Enter integer N0 (<101):'
read*,N0
N=4*N0

print*,'Enter integer N1 (<15):'
read*,N1
L=N1**2
NL=2*N+L

print*,'Enter time-step dt:'
read*,dt
tau=0.25d0

pi=4d0*datan(1d0)
alpha=1d0

dl=1d0/dfloat(N0)

do 10 i=1,N0
```

```fortran
      xb(i)=dfloat(i-1)*dl
      yb(i)=0d0
      xb(N0+i)=1d0
      yb(N0+i)=xb(i)
      xb(2*N0+i)=1d0-xb(i)
      yb(2*N0+i)=1d0
      xb(3*N0+i)=0d0
      yb(3*N0+i)=xb(2*N0+i)
10    continue

      xb(N+1)=xb(1)
      yb(N+1)=yb(1)

      do 20 i=1,N
      xm(i)=xb(i)+tau*(xb(i+1)-xb(i))
      ym(i)=yb(i)+tau*(yb(i+1)-yb(i))
      xm(N+i)=xb(i)+(1d0-tau)*(xb(i+1)-xb(i))
      ym(N+i)=yb(i)+(1d0-tau)*(yb(i+1)-yb(i))
      lg(i)=dsqrt((xb(i+1)-xb(i))**2d0
     & +(yb(i+1)-yb(i))**2d0)
      nx(i)=(yb(i+1)-yb(i))/lg(i)
      ny(i)=(xb(i)-xb(i+1))/lg(i)
20    continue

      dl=1d0/dfloat(N1+1)

      i=2*N

      do 25 j=1,N1
      do 25 k=1,N1
      i=i+1
      xm(i)=dfloat(j)*dl
      ym(i)=dfloat(k)*dl
25    continue

      do 26 i=1,N
      if ((i.le.N0).or.((i.gt.(2*N0))
```

```fortran
     & .and.(i.le.(3*NO)))) then
      BCT(i)=1
      else
      BCT(i)=0
      endif
26 continue

      do 27 i=1,NL
      phi(i)=1d0+dcos(0.25d0*pi*xm(i))
     & *dsin(0.25d0*pi*ym(i))
27 continue

      ti=-0.5d0*dt
      lud=1

28 ti=ti+dt
      tir=ti+0.5d0*dt

      do 30 i=1,N
      if (i.le.NO) then
      BCV(i)=-0.25d0*pi*dexp(-pi*pi*0.125d0*ti)
     & *dcos(0.25d0*pi*xm(i))
      BCV(N+i)=-0.25d0*pi*dexp(-pi*pi*0.125d0*ti)
     & *dcos(0.25d0*pi*xm(N+i))
      else if (i.le.(2*NO)) then
      BCV(i)=1d0+dexp(-pi*pi*0.125d0*tir)
     & *dsin(0.25d0*pi*ym(i))/dsqrt(2d0)
      BCV(N+i)=1d0+dexp(-pi*pi*0.125d0*tir)
     & *dsin(0.25d0*pi*ym(N+i))/dsqrt(2d0)
      else if (i.le.(3*NO)) then
      BCV(i)=0.25d0*pi*dexp(-pi*pi*0.125d0*ti)
     & *dcos(0.25d0*pi*xm(i))/dsqrt(2d0)
      BCV(N+i)=0.25d0*pi*dexp(-pi*pi*0.125d0*ti)
     & *dcos(0.25d0*pi*xm(N+i))/dsqrt(2d0)
      else
      BCV(i)=1d0+dexp(-pi*pi*0.125d0*tir)
     & *dsin(0.25d0*pi*ym(i))
```

```
      BCV(N+i)=1d0+dexp(-pi*pi*0.125d0*tir)
     & *dsin(0.25d0*pi*ym(N+i))
      endif
 30 continue

      call DLEDIFF(lud,N,L,tau,alpha,xm,ym,
     & xb,yb,nx,ny,lg,BCT,BCV,dt,phi)

      print*,'Time=',tir
      print*,' x y Numerical Exact'

      do 50 i=2*N+1,2*N+L
      write(*,60) xm(i),ym(i),phi(i),
     & 1d0+dexp(-pi*pi*0.125d0*tir)
     & *dcos(0.25d0*pi*xm(i))*dsin(0.25d0*pi*ym(i))
 50 continue

 60 format(4F14.6)

      print*,'To continue with the next
     & time level enter 1:'
      read*,ians
      lud=0

      if (ians.eq.1) goto 28

      end
```

The subprograms needed to compile **EX4PT1** into an executable code are the functions **CHI**, **DCHI** and **RHO** and the subroutines **DLEDIFF**, **DFMU**, **DPF** and **SOLVER** (with its supporting subprograms).

The problem under consideration is solved numerically as coded in **EX4PT1**. In Table 4.1, the numerical values of ϕ obtained using $N_0 = 10$, $N_1 = 9$ and $\Delta t = 0.10$ (with $\tau = 0.25$ for the discontinuous linear elements) are compared with the exact solution at the interior collocation points at $t = 1.0$. The numerical and the exact values are in reasonable agreement. The corresponding numerical values obtained using $N_0 = 20$, $N_1 = 7$ and $\Delta t = 0.05$ (also with $\tau = 0.25$) are also given in Table 4.1. We observe a significant improvement in the accuracy of the numerical values when the calculation is refined using more boundary elements and collocation points and a smaller time-step.

Table 4.1

(x, y)	$N_0 = 10$ $N_1 = 3$ $\Delta t = 0.10$	$N_0 = 20$ $N_1 = 7$ $\Delta t = 0.05$	Exact
$(0.25, 0.25)$	1.055768	1.055731	1.055721
$(0.25, 0.50)$	1.109297	1.109299	1.109301
$(0.25, 0.75)$	1.158632	1.158669	1.158681
$(0.50, 0.25)$	1.052541	1.052501	1.052488
$(0.50, 0.50)$	1.102952	1.102957	1.102959
$(0.50, 0.75)$	1.149416	1.149459	1.149474
$(0.75, 0.25)$	1.047282	1.047248	1.047238
$(0.75, 0.50)$	1.092657	1.092659	1.092661
$(0.75, 0.75)$	1.134478	1.134512	1.134523

Example 4.2

Repeat the problem in Example 4.1 with the initial-boundary conditions

$$\phi = x(1 - x)\sin(\pi y) \text{ at } t = 0 \text{ for } 0 < x < 1,\, 0 < y < 1,$$
$$\phi = 0 \text{ on } x = 0 \text{ for } 0 < y < 1,\, t > 0,$$
$$\phi = 0 \text{ on } x = 1 \text{ for } 0 < y < 1,\, t > 0,$$
$$\phi = 0 \text{ on } y = 0 \text{ for } 0 < x < 1,\, t > 0,$$
$$\phi = 0 \text{ on } y = 1 \text{ for } 0 < x < 1,\, t > 0.$$

The exact solution is available in series form as given by[1]

$$\phi = 4 \sum_{m=1}^{\infty} \frac{[1 - \cos(m\pi)]}{\pi^3 m^3} \sin(m\pi x) \sin(\pi y) \exp(-[(m\pi)^2 + \pi^2]t).$$

The boundary elements and the collocation points are chosen as explained in Example 4.1 above. The subroutine DLEDIFF is used to solve numerically the problem with $h(x, y) = x(1 - x) \sin(\pi y)$. Numerical values of ϕ at the interior point $(0.60, 0.60)$, as obtained by using $N_0 = 20$, $N_1 = 4$, $\tau = 0.25$ and $\Delta t = 0.01$, are compared with the exact solution in Table 4.2 at various time level. When the calculation is repeated using more boundary elements and collocation points with $N_0 = 40$ and $N_1 = 9$ and a smaller time-step with $\Delta t = 0.005$, the numerical values clearly converge to the exact ones. For this particular problem, ϕ decays rapidly to zero as time t increases. Hence, to obtain reasonably accurate numerical results such as those in Table 4.2, the time-step Δt in the calculation has to be quite small compared with that used in Example 4.1.

Table 4.2

t	$N_0 = 20$ $N_1 = 4$ $\Delta t = 0.01$	$N_0 = 40$ $N_1 = 9$ $\Delta t = 0.005$	Exact
0.01	0.189730	0.189572	0.189582
0.03	0.128526	0.128755	0.128808
0.05	0.086515	0.086867	0.086942
0.07	0.058178	0.058530	0.058604
0.09	0.039117	0.039427	0.039492
0.10	0.032074	0.032359	0.032418
0.20	0.004407	0.004487	0.004503
0.30	0.000605	0.000622	0.000626
0.40	0.000083	0.000086	0.000087

[1]This may be worked out using the formulae on pages 316 and 317 of the book *Partial Differential Equations in Mechanics 1* by APS Selvadurai (Springer-Verlag, 2000).

4.6 Summary and Discussion

The integro-differential equation in Eq. (4.4) is used to obtain a dual-reciprocity boundary element method for solving the initial-boundary value problem defined by Eqs. (4.1) and (4.2).

The method uses radial basis functions to approximate $\partial\phi/\partial t$ in order to reduce the double integral in Eq. (4.4) to a line integral over the boundary C of the spatial domain R. The boundary C is discretized into straight line elements and ϕ and $\partial\phi/\partial n$ are approximated as varying linearly across each element. Eq. (4.4) is then approximately reduced to a system of linear algebraic-differential equations containing unknown functions of time and first order time derivatives of some of these functions. The first order time derivatives are approximated using an accurate finite difference formula in order to obtain a time-stepping scheme for the approximate solution of the linear algebraic-differential equations.

The initial-boundary value problem is eventually reduced to a linear algebraic system $\mathbf{AX} = \mathbf{B}$ to be solved at consecutive time levels. At each time level, the values of either ϕ or $\partial\phi/\partial n$ at selected points on the boundary elements are among the unknowns to be determined. Although the dual-reciprocity boundary element formulation does not require the spatial domain R to be discretized into small cells (in the treatment of the double integral), the values of ϕ at selected collocation points in the interior of R appear as unknowns to be determined. The matrix $\mathbf{A}$ does not change with time. It is dependent only on the time-step between two consecutive time levels. Thus, as long as the time-step between two consecutive time levels and the boundary elements and interior collocation points used remain the same, the matrix $\mathbf{A}$ has to be evaluated and processed only once.

There are other boundary element approaches for solving Eqs. (4.1) and (4.2). If we take the Laplace transform of Eq. (4.1) and the boundary conditions in Eq. (4.2) with respect to time t, we obtain

$$\frac{\partial^2 \psi}{\partial x^2} + \frac{\partial^2 \psi}{\partial y^2} = \alpha[p\psi - f_0(x,y)], \qquad (4.18)$$

and

$$\psi(x,y,p) = F_1(x,y,p) \text{ for } (x,y) \in C_1,$$
$$\frac{\partial}{\partial n}[\psi(x,y,p)] = F_2(x,y,p) \text{ for } (x,y) \in C_2, \qquad (4.19)$$

where p is the Laplace transform parameter, $\psi(x,y,p)$, $F_1(x,y,p)$ and $F_2(x,y,p)$ are the Laplace transforms of $\phi(x,y,t)$, $f_1(x,y,t)$ and $f_2(x,y,t)$ respectively and $f_0(x,y)$ is the function in the initial condition given in Eq. (4.2).

Eq. (4.18) is essentially the Helmholtz type equation considered in Chapter 3 and may be solved by using a dual-reciprocity boundary element approach. Once ψ is obtained for selected values of p, a numerical method for inverting Laplace transform[2] may be employed to recover ϕ.

4.7   Exercises

1. Use the subroutine DLEDIFF for solving numerically Eq. (4.1) with $\alpha = 1/2$ within the triangular region R defined by $y < x$, $x > 0$, $y > 0$, subject to the initial-boundary conditions

$$\phi(x,y,0) = \exp(-\frac{1}{2}x)\cos(\frac{1}{2}x) - \exp(-\frac{1}{2}y)\cos(\frac{1}{2}y)$$
$$\text{for } (x,y) \in R,$$
$$\phi(x,y,t) = 0 \text{ for } y = x, \ 0 < x < 1, \ t > 0,$$

$$\frac{\partial \phi}{\partial n} = \frac{1}{2}[\cos(t) - \sin(t)] \text{ for } x = 0, \ 0 < y < 1, \ t > 0,$$
$$\frac{\partial \phi}{\partial n} = -\frac{1}{2}[\cos(t) - \sin(t)] \text{ for } y = 0, \ 0 < x < 1, \ t > 0.$$

---

[2] A useful numerical method for inverting Laplace transform is described in the paper by H Stehfest, "Numerical inversion of the Laplace transform" in *Communications of ACM* (Volume 13, 1970, pages 47-49 and 624).

Compare your numerical results with the exact solution given by

$$\phi(x, y, t) = \exp(-\frac{1}{2}x)\cos(t - \frac{1}{2}x)$$
$$- \exp(-\frac{1}{2}y)\cos(t - \frac{1}{2}y).$$

2. Use DLEDIFF to solve Eq. (4.2) in the square region $0 < x < 1, 0 < y < 1$, subject to

$$\begin{aligned}
\phi &= 1 \text{ at } t = 0 \text{ for } 0 < x < 1, 0 < y < 1, \\
\phi &= 0 \text{ on } x = 0 \text{ for } 0 < y < 1, t > 0, \\
\phi &= 0 \text{ on } x = 1 \text{ for } 0 < y < 1, t > 0, \\
\phi &= 0 \text{ on } y = 0 \text{ for } 0 < x < 1, t > 0, \\
\phi &= 0 \text{ on } y = 1 \text{ for } 0 < x < 1, t > 0.
\end{aligned}$$

(Note. In general, this problem may be quite challenging from a numerical point of view. There is a discontinuity in the initial-boundary conditions. At $t = 0$, ϕ is 0 on the boundary of the square region but equals to 1 at interior points. Thus, one may expect $\partial\phi/\partial n$ to be undefined on the boundary at $t = 0$. Fortunately, the boundary value of $\partial\phi/\partial n$ at $t = 0$ is not required by DLEDIFF. The real variables phi(1:2*N) and phi(2*N+1:2*N+L) should be assigned the values 0 and 1 respectively before DLEDIFF is called for the first time. If the discretization of the boundary is not sufficiently refined, the number of collocation points used is not large enough and the time-step is not small enough, then the numerical results obtained may be grossly inaccurate, especially at small time t. If the boundary is discretized into elements of equal length and the interior collocation points are employed as in Example 4.1 above, then $N_0 = 40$, $N_1 = 10$ and $\Delta t = 0.0050$ with $\tau = 0.25$ should give quite an accurate numerical values of ϕ at the interior collocation points. The exact solution of this problem in series form is given by[3]

$$\phi = 4 \sum_{n=1}^{\infty} \sum_{m=1}^{\infty} \frac{[1 - \cos(m\pi)][1 - \cos(n\pi)]}{mn\pi^2}$$
$$\times \sin(m\pi x) \sin(n\pi y) \exp(-[(m\pi)^2 + (n\pi)^2]t).$$

To evaluate the exact solution accurately for small t, one may have to retain a large number of terms in the series.)

---

[3]This is given on page 317 of the book *Partial Differential Equations in Mechanics 1* by APS Selvadurai (Springer-Verlag, 2000).

Chapter 5

Green's Functions for Potential Problems

5.1 Introduction

In Chapter 1, for the two-dimensional Laplace's equation, we have derived the boundary integral equation

$$\lambda(\xi,\eta)\phi(\xi,\eta) \;=\; \int_C [\phi(x,y)\frac{\partial}{\partial n}(\Phi(x,y;\xi,\eta))$$

$$-\Phi(x,y;\xi,\eta)\frac{\partial}{\partial n}(\phi(x,y))]ds(x,y). \quad (5.1)$$

Here ϕ satisfies the Laplace's equation in the two-dimensional region R bounded by a simple closed curve C on the Oxy plane, Φ is the fundamental solution given by

$$\Phi(x,y;\xi,\eta) = \frac{1}{4\pi}\ln([x-\xi]^2+[y-\eta]^2), \quad (5.2)$$

and λ is the parameter defined by

$$\lambda(\xi,\eta) = \begin{cases} 0 & \text{if } (\xi,\eta) \notin R \cup C, \\ 1/2 & \text{if } (\xi,\eta) \text{ lies on a smooth part of } C, \\ 1 & \text{if } (\xi,\eta) \in R. \end{cases} \quad (5.3)$$

The boundary integral equation in Eq. (5.1) may still be valid with possibly only minor modification of Eq. (5.3), if

$\Phi(x, y; \xi, \eta)$ is chosen to take the more general form

$$\Phi(x, y; \xi, \eta) = \frac{1}{4\pi} \ln([x - \xi]^2 + [y - \eta]^2) + \Phi^*(x, y; \xi, \eta), \quad (5.4)$$

where $\Phi^*(x, y; \xi, \eta)$ is any well defined function satisfying

$$\frac{\partial^2}{\partial x^2}[\Phi^*(x, y; \xi, \eta)] + \frac{\partial^2}{\partial y^2}[\Phi^*(x, y; \xi, \eta)] = 0$$

$$\text{for any } (x, y) \text{ and } (\xi, \eta) \text{ in } R. \quad (5.5)$$

Instead of taking $\Phi^* = 0$ (like in Chapters 1 and 2), we may find it advantageous to choose Φ^* that satisfies certain boundary conditions. The function Φ in Eq. (5.4) with a specially chosen Φ^* may be referred to as a Green's function. As we shall see later on, if an appropriately chosen Green's function, instead of the usual fundamental solution in Eq. (5.2), is used in the boundary integral equation, it may be possible to avoid integration over part of the boundary C. The number of boundary elements needed in the discretization of the boundary integral equation (hence the number of unknowns in the resulting system of linear algebraic equations) may then be reduced[1].

In this chapter, boundary element solutions of particular potential problems, obtained by using Green's functions for some special domains and boundary conditions, are presented with examples of applications.

5.2 Half Plane

5.2.1 Two Special Green's Functions

For the half plane $y > 0$ in Figure 5.1, two well known Green's functions satisfying particular conditions on the boundary $y =$

---

[1]An example of potential problems solved using a boundary integral formulation with Green's function, may be found in the article "A method for the numerical solution of some elliptic boundary value problems for a strip" by DL Clements and J Crowe in the *International Journal of Computer Mathematics* (Volume 8, 1980, pp. 345-355). The term "potential problems" refers to boundary value problems governed by the Laplace's equation.

0 are derived by the so called method of image. They are then applied to solve specific problems.

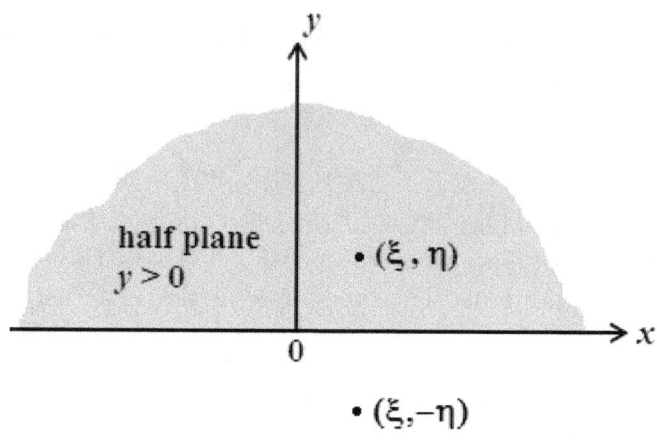

Figure 5.1

We choose $\Phi^*(x, y; \xi, \eta)$ such that it satisfies Eq. (5.5) in the half plane $y > 0$ and the boundary condition

$$\Phi(x, 0; \xi, \eta) = 0 \text{ for } -\infty < x < \infty, \tag{5.6}$$

which may be rewritten as

$$\Phi^*(x, 0; \xi, \eta) = -\frac{1}{4\pi} \ln([x - \xi]^2 + \eta^2) \text{ for } -\infty < x < \infty. \tag{5.7}$$

A suitable $\Phi^*(x, y; \xi, \eta)$ satisfying Eqs. (5.5) and (5.7) is given by

$$\Phi^*(x, y; \xi, \eta) = -\frac{1}{4\pi} \ln([x - \xi]^2 + [y + \eta]^2). \tag{5.8}$$

The function $\Phi^*(x, y; \xi, \eta)$ as given in Eq. (5.8) is well defined at all point (x, y) except at $(\xi, -\eta)$. If the point (ξ, η) is in the half plane then $(\xi, -\eta)$ being its image point obtained through reflection about $y = 0$ should be outside the half plane. Thus, $\Phi^*(x, y; \xi, \eta)$ in Eq. (5.8) satisfies the two-dimensional

Laplace's equation at all points (x, y) and (ξ, η) in the interior of the half plane.

Let us denote the Green's function given by Eqs. (5.4) and (5.8) by $\Phi_1(x, y; \xi, \eta)$, that is,

$$\Phi_1(x, y; \xi, \eta) = \frac{1}{4\pi} \ln([x - \xi]^2 + [y - \eta]^2)$$
$$- \frac{1}{4\pi} \ln([x - \xi]^2 + [y + \eta]^2). \quad (5.9)$$

For another Green's function, in place of the boundary condition in Eq. (5.6), $\Phi^*(x, y; \xi, \eta)$ is chosen to satisfy

$$\frac{\partial}{\partial y}(\Phi(x, y; \xi, \eta))\bigg|_{y=0} = 0 \quad \text{for} \quad -\infty < x < \infty, \quad (5.10)$$

that is,

$$\frac{\partial}{\partial y}(\Phi^*(x, y; \xi, \eta))\bigg|_{y=0} = \frac{\eta}{2\pi([x - \xi]^2 + \eta^2)} \quad \text{for} \quad -\infty < x < \infty. \quad (5.11)$$

It is easy to check that Eq. (5.11) is satisfied if

$$\Phi^*(x, y; \xi, \eta) = \frac{1}{4\pi} \ln([x - \xi]^2 + [y + \eta]^2). \quad (5.12)$$

We denote the Green's function given by Eqs. (5.4) and (5.12) by $\Phi_2(x, y; \xi, \eta)$. Thus,

$$\Phi_2(x, y; \xi, \eta) = \frac{1}{4\pi} \ln([x - \xi]^2 + [y - \eta]^2)$$
$$+ \frac{1}{4\pi} \ln([x - \xi]^2 + [y + \eta]^2). \quad (5.13)$$

We show now how the Green's functions in Eqs. (5.9) and (5.13) can be applied to obtain special boundary integral formulations for two particular cases of potential problems.

5.2.2 Applications

Case 5.1

This case involves a finite two-dimensional solution domain R which is a subset of the half plane $y > 0$. The curve boundary C of the region R consists of two non-intersecting parts denoted by D and E. The part D is an arbitrarily shaped open curve in the region $y > 0$, while E is a straight line segment on the x axis. For a geometrical sketch of the solution domain, refer to Figure 5.2.

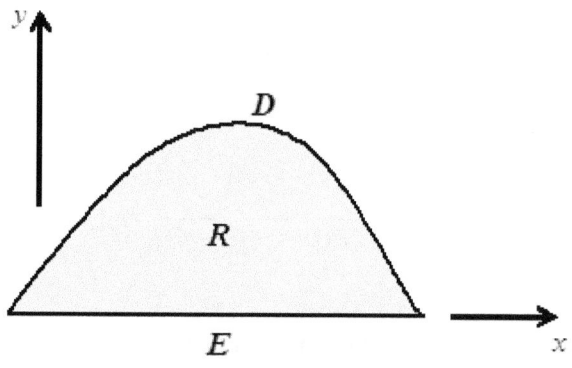

Figure 5.2

Mathematically, we are interested in solving the Laplace's equation

$$\frac{\partial^2 \phi}{\partial x^2} + \frac{\partial^2 \phi}{\partial y^2} = 0 \text{ in } R, \tag{5.14}$$

subject to

$$
\begin{aligned}
\phi &= 0 \text{ for } (x,y) \in E, \\
\phi &= f_1(x,y) \text{ for } (x,y) \in D_1, \\
\frac{\partial \phi}{\partial n} &= f_2(x,y) \text{ for } (x,y) \in D_2,
\end{aligned}
\tag{5.15}
$$

where f_1 and f_2 are suitably prescribed functions and D_1 and D_2 are non-intersecting curves such that $D_1 \cup D_2 = D$.

If we repeat the analysis in Section 1.4 (page 12, Chapter 1) using the Green's function $\Phi_1(x, y; \xi, \eta)$ in Eq. (5.9) in the place of the usual fundamental solution $\Phi(x, y; \xi, \eta) = (4\pi)^{-1}\ln([x - \xi]^2 + [y - \eta]^2)$, we obtain the boundary integral equation

$$
\begin{aligned}
\lambda(\xi, \eta)\phi(\xi, \eta) \;=\; & \int_D [\phi(x, y)\frac{\partial}{\partial n}(\Phi_1(x, y; \xi, \eta)) \\
& -\Phi_1(x, y; \xi, \eta)\frac{\partial}{\partial n}(\phi(x, y))]ds(x, y),
\end{aligned}
$$

$$(5.16)$$

with the parameter λ defined by

$$
\lambda(\xi, \eta) = \begin{cases} 0 & \text{if } (\xi, \eta) \in E, \\ 1/2 & \text{if } (\xi, \eta) \text{ lies on a smooth part of } D, \\ 1 & \text{if } (\xi, \eta) \in R. \end{cases}
$$

$$(5.17)$$

Note that the integral over E does not appear in Eq. (5.16). The conditions that $\phi(x, y)$ and $\Phi_1(x, y; \xi, \eta)$ are both zero for $(x, y) \in E$ are applied in the derivation of Eq. (5.16).

There is no need to discretize the boundary E if Eq. (5.16) is used to derive a boundary element procedure for solving the boundary value problem defined by Eqs. (5.14) and (5.15).

To discretize the integral in Eq. (5.16), let us put $N+1$ well spaced out points $(x^{(1)}, y^{(1)})$, $(x^{(2)}, y^{(2)})$, $\cdots$, $(x^{(N-1)}, y^{(N-1)})$, $(x^{(N)}, y^{(N)})$ and $(x^{(N+1)}, y^{(N+1)})$ on the open curve D. The end points of the curve D are $(x^{(1)}, y^{(1)})$ and $(x^{(N+1)}, y^{(N+1)})$ with $x^{(N+1)} < x^{(1)}$. Note that $y^{(1)} = y^{(N+1)} = 0$. The points are arranged such that $(x^{(k)}, y^{(k)})$ and $(x^{(k+1)}, y^{(k+1)})$ ($k = 1, 2, \cdots, N$) are two consecutive neighboring points. The line segment between $(x^{(k)}, y^{(k)})$ and $(x^{(k+1)}, y^{(k+1)})$ forms the element $D^{(k)}$. The curve D is approximated using

$$D \simeq D^{(1)} \cup D^{(2)} \cup \cdots \cup D^{(N-1)} \cup D^{(N)}. \qquad (5.18)$$

For constant elements, we may make the approximation

$$\phi \simeq \overline{\phi}^{(k)} \text{ and } \frac{\partial\phi}{\partial n} \simeq \overline{p}^{(k)} \text{ for } (x, y) \in D^{(k)} \ (k = 1, 2, \cdots, N),$$

$$(5.19)$$

161

in order to rewrite Eq. (5.16) approximately as

$$\lambda(\xi,\eta)\phi(\xi,\eta) \simeq \sum_{k=1}^{N}\overline{\phi}^{(k)}\int_{D^{(k)}}\frac{\partial}{\partial n}(\Phi_1(x,y;\xi,\eta))ds(x,y)$$

$$-\sum_{k=1}^{N}\overline{p}^{(k)}\int_{D^{(k)}}\Phi_1(x,y;\xi,\eta)ds(x,y).\text{(5.20)}$$

Proceeding as before, we may choose (ξ,η) in Eq. (5.20) to be given in turn by each of the midpoints of the boundary elements to set up a system of N linear algebraic equations to solve for either $\overline{\phi}^{(k)}$ or $\overline{p}^{(k)}$ (whichever is unknown on $D^{(k)}$). For setting up the linear algebraic equations, the integrals over $D^{(k)}$ can be evaluated analytically by

$$\int_{D^{(k)}}\Phi_1(x,y;\xi,\eta)ds(x,y) = \mathcal{F}_1^{(k)}(\xi,\eta) - \mathcal{F}_1^{(k)}(\xi,-\eta),$$

$$\int_{D^{(k)}}\frac{\partial}{\partial n}[\Phi_1(x,y;\xi,\eta)]ds(x,y) = \mathcal{F}_2^{(k)}(\xi,\eta) - \mathcal{F}_2^{(k)}(\xi,-\eta),$$

$$(5.21)$$

where the formulae for $\mathcal{F}_1^{(k)}(\xi,\eta)$ and $\mathcal{F}_2^{(k)}(\xi,\eta)$ are given in Section 1.6 (page 22, Chapter 1).

With Eq. (5.21), the subroutines CELAP1 and CELAP2 in Section 1.7 (page 28, Chapter 1) can be easily modified for the problem under consideration here to find unknown values of ϕ or $\partial\phi/\partial n$ on the elements and to compute ϕ at interior points in R. We modify CELAP1 and CELAP2 to create the subroutines G1LAP1 and G1LAP2 as listed below. In G1LAP1 or G1LAP2, the subroutine CPF is called to compute $\mathcal{F}_1^{(k)}(\xi,\eta)$ first and then again to calculate $\mathcal{F}_1^{(k)}(\xi,-\eta)$. The input and output parameters of G1LAP1 or G1LAP2 are exactly the same as those of CELAP1 and CELAP2 respectively, as detailed in Section 1.7.

```
      subroutine G1LAP1(N,xm,ym,xb,yb,nx,ny,lg,
     & BCT,BCV,phi,dphi)
```

```
      integer m,k,N,BCT(1000)

      double precision xm(1000),ym(1000),
     & xb(1000),yb(1000),nx(1000),ny(1000),
     & lg(1000),BCV(1000),A(1000,1000),B(1000),
     & pi,P1F1,P1F2,del,phi(1000),dphi(1000),
     & F1,F2,Z(1000),P2F1,P2F2

      pi=4d0*datan(1d0)

      do 10 m=1,N
      B(m)=0d0
      do 5 k=1,N
      call CPF(xm(m),ym(m),xb(k),yb(k),nx(k),ny(k),
     & lg(k),P1F1,P1F2)
      call CPF(xm(m),-ym(m),xb(k),yb(k),nx(k),ny(k),
     & lg(k),P2F1,P2F2)
      F1=(P1F1-P2F1)/pi
      F2=(P1F2-P2F2)/pi
      if (k.eq.m) then
      del=1d0
      else
      del=0d0
      endif
      if (BCT(k).eq.0) then
      A(m,k)=-F1
      B(m)=B(m)+BCV(k)*(-F2+0.5d0*del)
      else
      A(m,k)=F2-0.5d0*del
      B(m)=B(m)+BCV(k)*F1
      endif
 5    continue
 10   continue

      call solver(A,B,N,1,Z)
```

```fortran
      do 15 m=1,N
      if (BCT(m).eq.0) then
      phi(m)=BCV(m)
      dphi(m)=Z(m)
      else
      phi(m)=Z(m)
      dphi(m)=BCV(m)
      endif
15 continue

      return
      end

      subroutine G1LAP2(N,xi,eta,xb,yb,
     & nx,ny,lg,phi,dphi,pint)

      integer N,i

      double precision xi,eta,xb(1000),yb(1000),
     & nx(1000),ny(1000),lg(1000),phi(1000),
     & dphi(1000),pint,sum,pi,P1F1,P1F2,
     & P2F1,P2F2

      pi=4d0*datan(1d0)
      sum=0d0

      do 10 i=1,N
      call CPF(xi,eta,xb(i),yb(i),nx(i),ny(i),
     & lg(i),P1F1,P1F2)
      call CPF(xi,-eta,xb(i),yb(i),nx(i),ny(i),
     & lg(i),P2F1,P2F2)
      sum=sum+phi(i)*(P1F2-P2F2)-dphi(i)*(P1F1-P2F1)
10 continue

      pint=sum/pi

      return
```

```
end
```

Example 5.1

To test the subroutines G1LAP1 or G1LAP2, we use them to
solve Eq. (5.14) numerically in the square region $0 < x < 1$,
$0 < y < 1$, subject to

$$\phi(x,0) = 0 \text{ for } 0 < x < 1,$$
$$\frac{\partial \phi}{\partial n}\bigg|_{x=0} = 0 \text{ and } \frac{\partial \phi}{\partial n}\bigg|_{x=1} = 0 \text{ for } 0 < y < 1,$$
$$\phi(x,1) = 4x(1-x) \text{ for } 0 < x < 1.$$

The analytical solution for this particular problem is given
in series form by

$$\phi(x,y) = \frac{2}{3}y - 4\sum_{n=1}^{\infty} \frac{\sinh(2n\pi y)\cos(2n\pi x)}{n^2\pi^2 \sinh(2n\pi)}.$$

To use G1LAP1 or G1LAP2, we have to discretize only the
vertical sides of the square domain and the horizontal side $0 <
x < 1$, $y = 1$ into elements. Each of the sides is discretized
into N_0 equal length elements. Thus, $N = 3N_0$ and the points
$(x^{(1)}, y^{(1)})$, $(x^{(2)}, y^{(2)})$, $\cdots$, $(x^{(N-1)}, y^{(N-1)})$, $(x^{(N)}, y^{(N)})$ and
$(x^{(N+1)}, y^{(N+1)})$ (on the three sides) are given by

$$\left.\begin{array}{l} (x^{(k)}, y^{(k)}) = (1, [k-1]/N_0) \\ (x^{(N_0+k)}, y^{(N_0+k)}) = (1 - [k-1]/N_0, 1) \\ (x^{(2N_0+k)}, y^{(2N_0+k)}) = (0, 1 - [k-1]/N_0) \end{array}\right\} 1 \le k \le N_0,$$
$$(x^{(3N_0+1)}, y^{(3N_0+1)}) = (0,0).$$

The main program EX5PT1 which makes use of G1LAP1 and
G1LAP2 to solve the problem under consideration is listed below.

```
program EX5PT1

integer N0,BCT(1000),N,i,ians
```

```
      double precision xb(1000),yb(1000),xm(1000),
     & ym(1000),nx(1000),ny(1000),lg(1000),BCV(1000),
     & phi(1000),dphi(1000),pint,dl,xi,eta,pi,exct

      print*,'Enter number of elements per side
     & (<334):'
      read*,N0
      N=3*N0

      pi=4d0*datan(1d0)
      dl=1d0/dfloat(N0)

      do 10 i=1,N0
      xb(i)=1d0
      yb(i)=dfloat(i-1)*dl
      xb(N0+i)=1d0-yb(i)
      yb(N0+i)=1d0
      xb(2*N0+i)=0d0
      yb(2*N0+i)=xb(N0+i)
 10   continue
      xb(N+1)=0d0
      yb(N+1)=0d0

      do 20 i=1,N
      xm(i)=0.5d0*(xb(i)+xb(i+1))
      ym(i)=0.5d0*(yb(i)+yb(i+1))
      lg(i)=dsqrt((xb(i+1)-xb(i))**2d0
     & +(yb(i+1)-yb(i))**2d0)
      nx(i)=(yb(i+1)-yb(i))/lg(i)
      ny(i)=(xb(i)-xb(i+1))/lg(i)
 20   continue

      do 30 i=1,N
      if (i.le.N0) then
      BCT(i)=1
      BCV(i)=0d0
```

```
      else if ((i.gt.N0).and.(i.le.(2*N0))) then
      BCT(i)=0
      BCV(i)=4d0*xm(i)*(1d0-xm(i))
      else
      BCT(i)=1
      BCV(i)=0d0
      endif
30 continue

      call G1LAP1(N,xm,ym,xb,yb,nx,ny,lg,
     & BCT,BCV,phi,dphi)

50 print*,'Enter coordinates xi and eta
     & of an interior point:'

      read*,xi,eta

      call G1LAP2(N,xi,eta,xb,yb,nx,ny,lg,
     & phi,dphi,pint)

      exct=2d0*eta/3d0
      do 55 i=1,1000
      exct=exct-4d0*((dexp(2d0*dfloat(i)*pi*(eta-1d0))
     & -dexp(-2d0*dfloat(i)*pi*(eta+1d0)))
     & *dcos(2d0*dfloat(i)*pi*xi))
     & /((dfloat(i*i)*pi*pi)
     & *(1d0-dexp(-4d0*dfloat(i)*pi)))
55 continue

      write(*,60)pint,exct
60 format('Numerical and exact values are:',
     & F14.6,' and',F14.6,' respectively')

      print*,'To continue with another point enter 1:'
      read*,ians

      if (ians.eq.1) goto 50
```

end

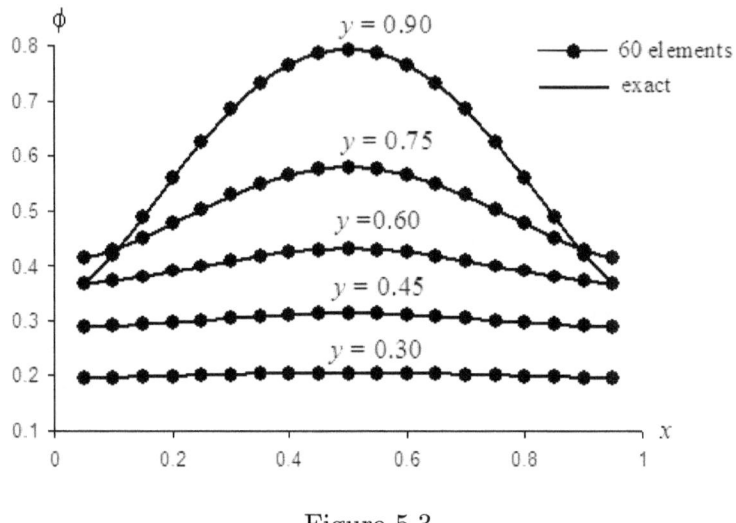

Figure 5.3

In Figure 5.3, we plot the numerical ϕ obtained using $N_0 = 20$ (60 boundary elements altogether) against $0 < x < 1$ for $y = 0.30, 0.45, 0.60, 0.75$ and 0.90. The graphs of the numerical and the exact solutions are in good agreement with each other.

Case 5.2

As sketched in Figure 5.4, the solution domain R is taken to be the half plane $y > 0$ without the finite region bounded by a simple closed curve C. We are interested in solving Eq. (5.14) subject to

$$\left.\frac{\partial \phi}{\partial n}\right|_{y=0} = 0 \text{ for } -\infty < x < \infty,$$
$$\phi = f_1(x, y) \text{ for } (x, y) \in C_1,$$
$$\frac{\partial \phi}{\partial n} = f_2(x, y) \text{ for } (x, y) \in C_2,$$
$$\phi \to 0 \text{ as } x^2 + y^2 \to \infty, \tag{5.22}$$

where f_1 and f_2 are suitably prescribed functions and C_1 and C_2 are non-intersecting curves such that $C_1 \cup C_2 = C$.

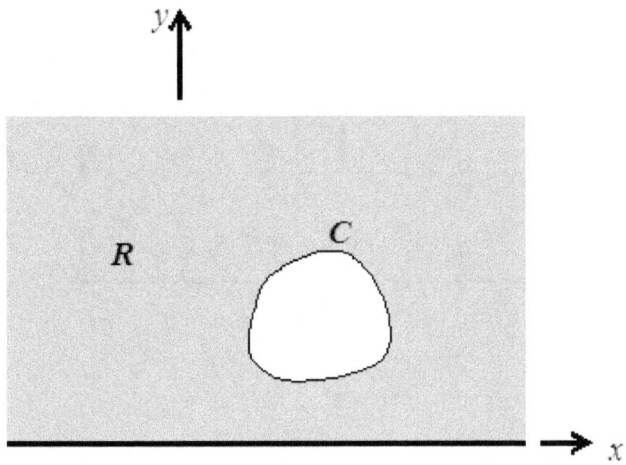

Figure 5.4

Note that $\partial\phi/\partial n = -\partial\phi/\partial y$ on the plane boundary $y = 0$. The last condition in Eq. (5.22) specifies the "far field" behavior of the solution. For our purpose here, we assume that ϕ decays as $O([x^2 + y^2]^{-a} \ln[x^2 + y^2])$ (with a being a positive real number) for large $x^2 + y^2$ (within the half plane).

Particular engineering problems which require the computation of stress around a hole or rigid inclusion or fluid flow past an impermeable body may be formulated in terms of the boundary value problem above. The curve C represents the boundary of the hole or rigid inclusion or impermeable body.

To obtain a boundary integral formulation for the problem under consideration, let us first introduce an artificial boundary S_ρ given by the semi-circle $x^2 + y^2 = \rho^2$, $y > 0$, where ρ is a positive real number. We take R_ρ to be the finite region whose boundary is given by $C \cup S_\rho \cup L_\rho$, where L_ρ is the horizontal straight line between the point $(-\rho, 0)$ and $(\rho, 0)$. We assume that ρ is sufficiently large, so that the simple closed curve C lies wholly in the region R_ρ. Refer to Figure 5.5. We can recover

169

the solution domain R in Figure 5.4 if we let the parameter ρ tend to infinity.

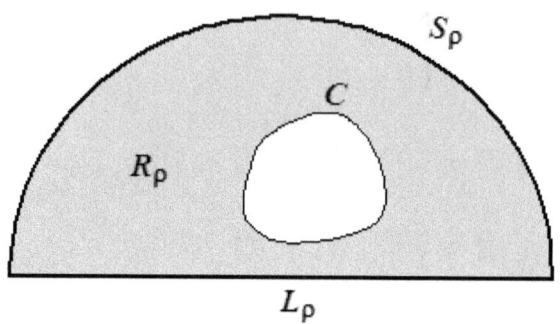

Figure 5.5

If we carry out the analysis in Section 1.4 (page 12, Chapter 1) on the region R_ρ using the Green's function $\Phi_2(x, y; \xi, \eta)$ in Eq. (5.13), we obtain the boundary integral equation

$$\lambda(\xi, \eta)\phi(\xi, \eta) = \int_{C \cup S_\rho} [\phi(x, y)\frac{\partial}{\partial n}(\Phi_2(x, y; \xi, \eta))$$

$$-\Phi_2(x, y; \xi, \eta)\frac{\partial}{\partial n}(\phi(x, y))]ds(x, y),$$

$$(5.23)$$

with the parameter λ defined by

$$\lambda(\xi, \eta) = \begin{cases} 1/2 & \text{if } (\xi, \eta) \text{ lies on a smooth part of } C \cup S_\rho, \\ 1 & \text{if } (\xi, \eta) \in R_\rho \cup L_\rho. \end{cases}$$

$$(5.24)$$

In deriving Eq. (5.23), we use the boundary conditions $\partial\phi/\partial n = 0$ and $\partial\Phi_2/\partial n = 0$ on L_ρ. The integral over L_ρ does not appear in the formulation.

To examine the integral over S_ρ for ρ tending to infinity, we write

$$\begin{aligned}
&\ln([x-\xi]^2 + [y \pm \eta]^2) \\
&= \ln(r^2 + \xi^2 + \eta^2 - 2r[\xi\cos\theta \mp \eta\sin\theta]) \\
&= \ln(r^2) + \ln(1 + \frac{\xi^2 + \eta^2}{r^2} - \frac{2[\xi\cos\theta \mp \eta\sin\theta]}{r}),
\end{aligned}$$

$$(5.25)$$

where r and θ are the usual polar coordinates.

From Eq. (5.25), it is obvious that $\ln([x-\xi]^2 + [y \pm \eta]^2) \simeq \ln(r^2)$ for large r. It follows that

$$\Phi_2 \simeq \frac{1}{\pi}\ln(r) \quad \text{for large } r.$$

If we further assume that

$$\phi \simeq A(\theta)\frac{\ln(r)}{r^{2a}} \quad \text{for large } r,$$

where $A(\theta)$ is a well defined function of θ and a is a positive real number, we find that

$$\begin{aligned}
&\lim_{\rho \to \infty} \int_{S_\rho} [\phi(x,y)\frac{\partial}{\partial n}(\Phi_2(x,y;\xi,\eta)) \\
&\quad -\Phi_2(x,y;\xi,\eta)\frac{\partial}{\partial n}(\phi(x,y))]ds(x,y) \\
&= \lim_{\rho \to \infty} \frac{[\ln(\rho)]^2}{\pi\rho^{2a}} \int_0^\pi 2aA(\theta)d\theta = 0.
\end{aligned}$$

Thus, if we let ρ tend to infinity in Eq. (5.23), for the potential problem here with conditions given in Eq. (5.22), we obtain the boundary integral equation

$$\begin{aligned}
\lambda(\xi,\eta)\phi(\xi,\eta) &= \int_C [\phi(x,y)\frac{\partial}{\partial n}(\Phi_2(x,y;\xi,\eta)) \\
&\quad -\Phi_2(x,y;\xi,\eta)\frac{\partial}{\partial n}(\phi(x,y))]ds(x,y),
\end{aligned}$$

$$(5.26)$$

171

with the parameter λ defined by

$$\lambda(\xi, \eta) = \begin{cases} 1/2 & \text{if } (\xi, \eta) \text{ lies on a smooth part of } C, \\ 1 & \text{if } (\xi, \eta) \in R \text{ or if } \eta = 0. \end{cases}$$

(5.27)

Note that in Eq. (5.26) the integral is over the boundary C only. If the usual fundamental solution $\Phi(x, y; \xi, \eta) = (4\pi)^{-1} \ln([x - \xi]^2 + [y - \eta]^2)$ is used instead of $\Phi_2(x, y; \xi, \eta)$, the path of integration in the boundary integral equation includes the line $y = 0$ for $-\infty < x < \infty$. The advantage of using the special Green's function $\Phi_2(x, y; \xi, \eta)$ is obvious for the particular problem under consideration.

Finally, let us check that the integral expression for $\phi(\xi, \eta)$ in Eq. (5.26) tends to zero as $\xi^2 + \eta^2 \to \infty$ (within the solution domain). From Eq. (5.13), for $(x, y) \in C$, we find that $\Phi_2(x, y; \xi, \eta) \to (2\pi)^{-1} \ln(\xi^2 + \eta^2)$ and $\partial(\Phi_2(x, y; \xi, \eta)/\partial n \to 0$ as $\xi^2 + \eta^2 \to \infty$. Thus, Eq. (5.26) gives

$$\phi(\xi, \eta) \quad \to \quad -\frac{1}{2\pi} \ln(\xi^2 + \eta^2) \int_C \frac{\partial}{\partial n}(\phi(x, y)) ds(x, y)$$

$$\text{as } \xi^2 + \eta^2 \to \infty.$$

With reference to Figure 5.5, $\partial\phi/\partial n$ is required to satisfy[2]

$$\int_{C \cup L_\rho \cup S_\rho} \frac{\partial}{\partial n}(\phi(x, y)) ds(x, y) = 0.$$

Now we know that $\partial\phi/\partial n$ is zero on L_ρ. We expect $\partial\phi/\partial n$ to behave as $O([x^2 + y^2]^{-a-1/2} \ln[x^2 + y^2])$ on S_ρ for large ρ (since we assume that ϕ decays as $O([x^2 + y^2]^{-a} \ln[x^2 + y^2])$ for large $x^2 + y^2$). Thus, for the problem under consideration, if we let $\rho \to \infty$, we find that $\partial\phi/\partial n$ on C satisfies

$$\int_C \frac{\partial}{\partial n}(\phi(x, y)) ds(x, y) = 0.$$

---

[2]Exercise 1 of Chapter 1 on page 50 may be of relevance here.

It follows that $\phi(\xi, \eta)$ as given by Eq. (5.26) tends to zero as $\xi^2 + \eta^2 \to \infty$. Note that if we prescribe $\partial\phi/\partial n$ at all points on C we must ensure that the condition above is fulfilled.

Since the curve C is an inner boundary (within the half plane), the unit normal vector $[n_x, n_y]$ on C (for computing $\partial\phi/\partial n$ and $\partial\Phi_2/\partial n$) is taken to point into the region bounded by C. To discretize C, we place N well spaced out points $(x^{(1)}, y^{(1)}), (x^{(2)}, y^{(2)}), \cdots, (x^{(N-1)}, y^{(N-1)})$ and $(x^{(N)}, y^{(N)})$ on C in the clockwise (instead of counter clockwise) direction. Using the conditions specified on C and constant elements, we may use the computer codes in Chapter 1 to solve the problem under consideration. Since the Green's function Φ_2 is used here (in place of the usual fundamental solution Φ), the subroutines CELAP1 and CELAP2 for computing ϕ numerically have to be replaced by G2LAP1 and G2LAP2 as listed below.

```
    subroutine G2LAP1(N,xm,ym,xb,yb,nx,ny,lg,
& BCT,BCV,phi,dphi)

    integer m,k,N,BCT(1000)

    double precision xm(1000),ym(1000),,
& xb(1000),yb(1000)nx(1000),ny(1000),lg(1000),
& BCV(1000),A(1000,1000),B(1000),pi,P1F1,P1F2,
& del,phi(1000),dphi(1000),F1,F2,Z(1000),
& P2F1,P2F2

    pi=4d0*datan(1d0)

    do 10 m=1,N
    B(m)=0d0
    do 5 k=1,N
    call CPF(xm(m),ym(m),xb(k),yb(k),
& nx(k),ny(k),lg(k),P1F1,P1F2)
    call CPF(xm(m),-ym(m),xb(k),yb(k),
& nx(k),ny(k),lg(k),P2F1,P2F2)
    F1=(P1F1+P2F1)/pi
```

```
      F2=(P1F2+P2F2)/pi
      if (k.eq.m) then
      del=1d0
      else
      del=0d0
      endif
      if (BCT(k).eq.0) then
      A(m,k)=-F1
      B(m)=B(m)+BCV(k)*(-F2+0.5d0*del)
      else
      A(m,k)=F2-0.5d0*del
      B(m)=B(m)+BCV(k)*F1
      endif
5   continue
10  continue

      call solver(A,B,N,1,Z)

      do 15 m=1,N
      if (BCT(m).eq.0) then
      phi(m)=BCV(m)
      dphi(m)=Z(m)
      else
      phi(m)=Z(m)
      dphi(m)=BCV(m)
      endif
15  continue

      return
      end

      subroutine G2LAP2(N,xi,eta,xb,yb,
     & nx,ny,lg,phi,dphi,pint)

      integer N,i

      double precision xi,eta,xb(1000),yb(1000),
```

```
& nx(1000),ny(1000),lg(1000),phi(1000),
& dphi(1000),pint,sum,pi,P1F1,P1F2,
& P2F1,P2F2

  pi=4d0*datan(1d0)
  sum=0d0

  do 10 i=1,N
  call CPF(xi,eta,xb(i),yb(i),nx(i),ny(i),
& lg(i),P1F1,P1F2)
  call CPF(xi,-eta,xb(i),yb(i),nx(i),ny(i),
& lg(i),P2F1,P2F2)
  sum=sum+phi(i)*(P1F2+P2F2)-dphi(i)*(P1F1+P2F1)
10 continue

  pint=sum/pi

  return
  end
```

Example 5.2

For a particular test problem, let us take the inner boundary C to be the circle $x^2 + (y-1)^2 = 1/4$. The boundary condition on the inner boundary is given by

$$\phi = 4x(1 + \frac{1}{16y+1}) \text{ on } C.$$

The condition on $y = 0$ and the far field condition are as described in Eq. (5.22).

It is easy to check that the exact solution of the particular test problem is given by

$$\phi = x(\frac{1}{x^2 + (y-1)^2} + \frac{1}{x^2 + (y+1)^2}).$$

The main program for numerical solution of the test problem is listed in **EX5PT2** below. Note that the boundary points on the circular boundary C are arranged in clockwise order.

175

```
      program EX5PT2

      integer BCT(1000),N,i,ians,j

      double precision xb(1000),yb(1000),
     & xm(1000),ym(1000),nx(1000),ny(1000),
     & lg(1000),BCV(1000),phi(1000),dphi(1000),
     & pint,dl,xi,eta,pi,exct

      print*,'Enter number of elements on the
     & circle (<1000):'
      read*,N

      pi=4d0*datan(1d0)
      dl=2d0*pi/dfloat(N)

      do 10 i=1,N
      xb(i)=0.5d0*dcos(dfloat(i-1)*dl)
      yb(i)=1d0-0.5d0*dsin(dfloat(i-1)*dl)
10    continue
      xb(N+1)=xb(1)
      yb(N+1)=yb(1)

      do 20 i=1,N
      xm(i)=0.5d0*(xb(i)+xb(i+1))
      ym(i)=0.5d0*(yb(i)+yb(i+1))
      lg(i)=dsqrt((xb(i+1)-xb(i))**2d0
     & +(yb(i+1)-yb(i))**2d0)
      nx(i)=(yb(i+1)-yb(i))/lg(i)
      ny(i)=(xb(i)-xb(i+1))/lg(i)
20    continue

      do 30 i=1,N
      BCT(i)=0
      BCV(i)=4d0*xm(i)
     & *(1d0+1d0/(16d0*ym(i)+1d0))
```

```
   30 continue

      call G2LAP1(N,xm,ym,xb,yb,nx,ny,
     & lg,BCT,BCV,phi,dphi)

   50 print*,'Enter coordinates xi and eta of
     & an interior point:'
      read*,xi,eta

      call G2LAP2(N,xi,eta,xb,yb,nx,ny,
     & lg,phi,dphi,pint)

      exct=xi*(1d0/(xi**2d0+(eta-1d0)**2d0)
     & +1d0/(xi**2d0+(eta+1d0)**2d0))

      write(*,60)pint,exct
   60 format('Numerical and exact values are:',
     & F14.6,' and',F14.6,' respectively')

      print*,'To continue with another point
     & enter 1:'
      read*,ians
      if (ians.eq.1) goto 50

      end
```

Discretizing the circular boundary C into 50 elements of equal length, we compute $\phi(x,0)$ numerically and compare the numerical values with the exact ones for $-5 < x < 5$ in Figure 5.6. This verifies that ϕ may be accurately calculated on the exterior boundary $y = 0$ of the solution domain by using Eq. (5.26) with $\lambda = 1$.

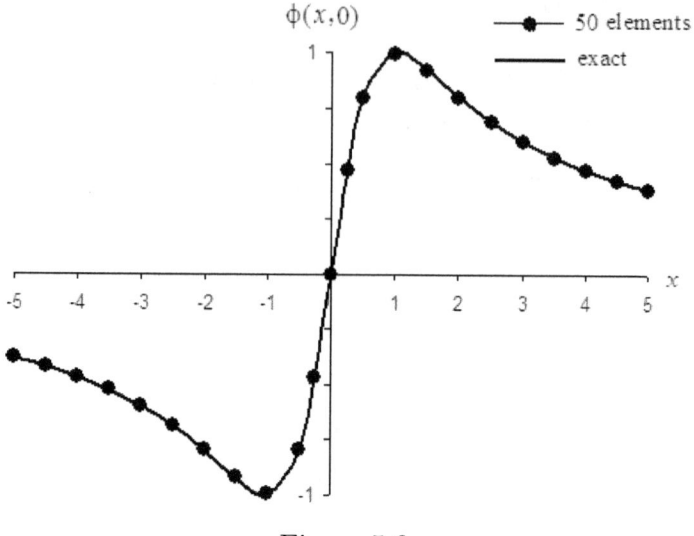

$\phi(x,0)$

— 50 elements

— exact

Figure 5.6

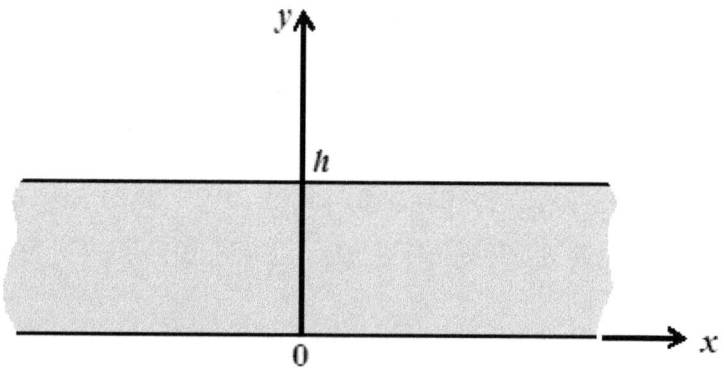

Figure 5.7

5.3   Infinitely Long Strip

5.3.1   Derivation of Green's Functions by Conformal Mapping

Let us consider the infinitely long strip $-\infty < x < \infty$, $0 < y < h$, on the Oxy plane, as shown in Figure 5.7. If we write $z = x + iy$ ($i = \sqrt{-1}$), the conformal mapping $w = \exp(\pi z/h)$

(with $w = u + iv$) can be used to transform the infinitely long strip to the half plane $v > 0$ (on the Ouv plane) in Figure 5.8. The boundary $y = 0$ is mapped to the line $v = 0$, $u > 0$, while $y = h$ to $v = 0$, $u < 0$.

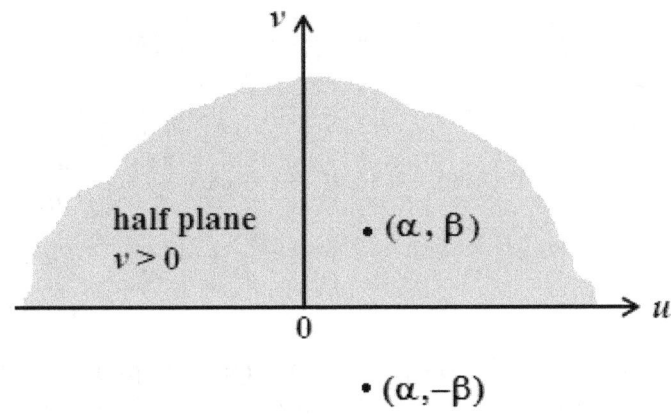

Figure 5.8

In real variables, the conformal mapping $w = \exp(\pi z/h)$ can be expressed in terms of a pair of equations given by

$$
\begin{aligned}
u &= \exp(\frac{\pi x}{h}) \cos(\frac{\pi y}{h}), \\
v &= \exp(\frac{\pi x}{h}) \sin(\frac{\pi y}{h}).
\end{aligned}
\tag{5.28}
$$

Let us construct a Green's function $\widetilde{\Psi}(u, v; \alpha, \beta)$ for the half plane $u > 0$ in Figure 5.8 such that $\widetilde{\Psi}$ is singular at $(u, v) = (\alpha, \beta)$ and satisfies the Dirichlet condition

$$
\widetilde{\Psi}(u, 0; \alpha, \beta) = 0 \text{ for } -\infty < u < \infty. \tag{5.29}
$$

From Eq. (5.9), such a Green's function is given by

$$
\begin{aligned}
\widetilde{\Psi}(u, v; \alpha, \beta) &= \frac{1}{4\pi} \ln([u - \alpha]^2 + [v - \beta]^2) \\
&\quad - \frac{1}{4\pi} \ln([u - \alpha]^2 + [v + \beta]^2).
\end{aligned}
\tag{5.30}
$$

If (α, β) is the image point of (ξ, η) (a given point in the infinitely long strip in Figure 5.7), we may use Eq. (5.28) to bring $\widetilde{\Psi}(u, v; \alpha, \beta)$ to the physical $0xy$ plane, that is, we define the function

$$
\begin{aligned}
\Psi(x&, y; \xi, \eta) \\
= \ & \frac{1}{4\pi} \ln([\exp(\frac{\pi x}{h}) \cos(\frac{\pi y}{h}) - \exp(\frac{\pi \xi}{h}) \cos(\frac{\pi \eta}{h})]^2 \\
& + [\exp(\frac{\pi x}{h}) \sin(\frac{\pi y}{h}) - \exp(\frac{\pi \xi}{h}) \sin(\frac{\pi \eta}{h})]^2) \\
& - \frac{1}{4\pi} \ln([\exp(\frac{\pi x}{h}) \cos(\frac{\pi y}{h}) - \exp(\frac{\pi \xi}{h}) \cos(\frac{\pi \eta}{h})]^2 \\
& + [\exp(\frac{\pi x}{h}) \sin(\frac{\pi y}{h}) + \exp(\frac{\pi \xi}{h}) \sin(\frac{\pi \eta}{h})]^2).
\end{aligned}
$$

$$(5.31)$$

According to the theory of conformal mapping[3], Ψ should satisfy the two-dimensional Laplace's equation everywhere in the infinitely long strip except at $(x, y) = (\xi, \eta)$ and the conditions

$$\Psi(x, 0; \xi, \eta) = 0 \ \text{ and } \ \Psi(x, h; \xi, \eta) = 0 \ \text{ for } -\infty < x < \infty.$$
$$(5.32)$$

Note that Eq. (5.32) follows directly from Eq. (5.29).

If R is the region bounded by a simple closed curve C which lies in the infinitely long strip, can we use the function $\Psi(x, y; \xi, \eta)$ in Eq. (5.31) in place of $\Phi(x, y; \xi, \eta)$ in Eq. (5.2) to derive the boundary integral equation in Eq. (5.1)?

To find out, let us examine what happens to $\Psi(x, y; \xi, \eta)$ as (x, y) approaches the point (ξ, η).

As (x, y) tends to (ξ, η), the first logarithmic term on the right hand side of Eq. (5.31) blows up but the second logarithmic term is bounded. Examining the argument inside the

---

[3]For further details, one may refer to the relevant chapters on conformal mapping in the textbook *Complex Variables and Applications* by RV Churchill and JW Brown (McGraw-Hill, 1990).

logarithmic function in the first term, we find that

$$\exp(\frac{\pi x}{h}) \cos(\frac{\pi y}{h}) - \exp(\frac{\pi \xi}{h}) \cos(\frac{\pi \eta}{h})$$

$$= (\exp(\frac{\pi \xi}{h}) + \frac{\pi}{h} \exp(\frac{\pi \xi}{h})[x - \xi]$$

$$+ \frac{\pi^2}{2h^2} \exp(\frac{\pi \xi}{h})[x - \xi]^2 + \cdots)$$

$$\times (\cos(\frac{\pi \eta}{h}) - \frac{\pi}{h} \sin(\frac{\pi \eta}{h})[y - \eta]$$

$$+ \frac{\pi^2}{2h^2} \cos(\frac{\pi \eta}{h})[y - \eta]^2 + \cdots)$$

$$- \exp(\frac{\pi \xi}{h}) \cos(\frac{\pi \eta}{h})$$

$$\simeq \frac{\pi}{h} \exp(\frac{\pi \xi}{h})([x - \xi] \cos(\frac{\pi \eta}{h}) - [y - \eta] \sin(\frac{\pi \eta}{h}))$$

for (x, y) very close to (ξ, η),

and

$$\exp(\frac{\pi x}{h}) \sin(\frac{\pi y}{h}) - \exp(\frac{\pi \xi}{h}) \sin(\frac{\pi \eta}{h})$$

$$= (\exp(\frac{\pi \xi}{h}) + \frac{\pi}{h} \exp(\frac{\pi \xi}{h})[x - \xi]$$

$$+ \frac{\pi^2}{2h^2} \exp(\frac{\pi \xi}{h})[x - \xi]^2 + \cdots)$$

$$\times (\sin(\frac{\pi \eta}{h}) + \frac{\pi}{h} \cos(\frac{\pi \eta}{h})[y - \eta]$$

$$- \frac{\pi^2}{2h^2} \sin(\frac{\pi \eta}{h})[y - \eta]^2 + \cdots)$$

$$- \exp(\frac{\pi \xi}{h}) \sin(\frac{\pi \eta}{h})$$

$$\simeq \frac{\pi}{h} \exp(\frac{\pi \xi}{h})([y - \eta] \cos(\frac{\pi \eta}{h}) + [x - \eta] \sin(\frac{\pi \eta}{h}))$$

for (x, y) very close to (ξ, η).

It follows that

$$\Psi(x, y; \xi, \eta) \simeq \frac{1}{4\pi} \ln([x - \xi]^2 + [y - \eta]^2)$$

for (x, y) very close to (ξ, η),

that is, $\Psi(x, y; \xi, \eta)$ behaves like the usual fundamental solution in Eq. (5.2) as (x, y) approaches (ξ, η).

Thus, if R is the region bounded by a simple closed curve C which lies in the infinitely long strip, we can use the function $\Psi(x, y; \xi, \eta)$ in Eq. (5.31) in place of $\Phi(x, y; \xi, \eta)$ in Eq. (5.2) to derive the boundary integral equation in Eq. (5.1).

For the infinitely long strip, let us define

$$
\begin{aligned}
&\Phi_3(x, y; \xi, \eta) \\
&= \frac{1}{4\pi} \ln([\exp(\frac{\pi x}{h}) \cos(\frac{\pi y}{h}) - \exp(\frac{\pi \xi}{h}) \cos(\frac{\pi \eta}{h})]^2 \\
&\quad + [\exp(\frac{\pi x}{h}) \sin(\frac{\pi y}{h}) - \exp(\frac{\pi \xi}{h}) \sin(\frac{\pi \eta}{h})]^2) \\
&\quad - \frac{1}{4\pi} \ln([\exp(\frac{\pi x}{h}) \cos(\frac{\pi y}{h}) - \exp(\frac{\pi \xi}{h}) \cos(\frac{\pi \eta}{h})]^2 \\
&\quad + [\exp(\frac{\pi x}{h}) \sin(\frac{\pi y}{h}) + \exp(\frac{\pi \xi}{h}) \sin(\frac{\pi \eta}{h})]^2), \qquad (5.33)
\end{aligned}
$$

as a Green's function (for the two-dimensional Laplace's equation) satisfying the conditions

$$
\Phi_3(x, 0; \xi, \eta) = 0 \ \text{ and } \ \Phi_3(x, h; \xi, \eta) = 0 \ \text{ for } \ -\infty < x < \infty.
\tag{5.34}
$$

Similarly, we may obtain

$$
\begin{aligned}
&\Phi_4(x, y; \xi, \eta) \\
&= \frac{1}{4\pi} \ln([\exp(\frac{\pi x}{h}) \cos(\frac{\pi y}{h}) - \exp(\frac{\pi \xi}{h}) \cos(\frac{\pi \eta}{h})]^2 \\
&\quad + [\exp(\frac{\pi x}{h}) \sin(\frac{\pi y}{h}) - \exp(\frac{\pi \xi}{h}) \sin(\frac{\pi \eta}{h})]^2) \\
&\quad + \frac{1}{4\pi} \ln([\exp(\frac{\pi x}{h}) \cos(\frac{\pi y}{h}) - \exp(\frac{\pi \xi}{h}) \cos(\frac{\pi \eta}{h})]^2 \\
&\quad + [\exp(\frac{\pi x}{h}) \sin(\frac{\pi y}{h}) + \exp(\frac{\pi \xi}{h}) \sin(\frac{\pi \eta}{h})]^2), \qquad (5.35)
\end{aligned}
$$

as a Green's function which satisfies the conditions[4]

$$\frac{\partial}{\partial y}[\Phi_4(x,y;\xi,\eta)]\Big|_{y=0} = 0 \text{ and } \frac{\partial}{\partial y}[\Phi_4(x,y;\xi,\eta)]\Big|_{y=h} = 0$$

$$\text{for } -\infty < x < \infty.$$

$$(5.36)$$

5.3.2 Applications

The Green's functions $\Phi_3(x,0;\xi,\eta)$ and $\Phi_4(x,y;\xi,\eta)$ above are applied to obtain special boundary integral formulations for two specific problems.

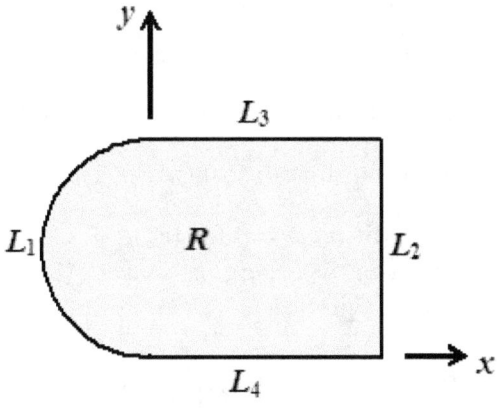

Figure 5.9

Example 5.3

Consider the solution domain R shown in Figure 5.9. The boundary of R comprises four parts denoted by L_1, L_2, L_3

---

[4]An alternative Green's function satisfying the conditions in Eq. (5.36), expressed in terms of a Fourier integral transform, may be found in the paper "A method for the numerical solution of some elliptic boundary value problems for a strip" by DL Clements and J Crowe in the *International Journal of Computer Mathematics* (Volume 8, 1980, pp. 345-355).

and L_4, where

$$
\begin{aligned}
L_1 &= \{(x,y): \ x^2 + (y - 1/2)^2 = 1/4, \ x < 0\}, \\
L_2 &= \{(x,y): \ x = 1, \ 0 < y < 1\}, \\
L_3 &= \{(x,y): \ 0 \le x < 1, \ y = 1\}, \\
L_4 &= \{(x,y): \ 0 \le x < 1, \ y = 0\}.
\end{aligned}
$$

The problem of interest here is to solve Eq. (5.14) in R subject to the boundary conditions

$$
\begin{aligned}
\frac{\partial \phi}{\partial n} &= 2\pi x \sin(\pi y) \sinh(\pi[x - 1]) \\
&\quad + \pi(2y - 1) \cos(\pi y) \cosh(\pi[x - 1]) \\
&\qquad \text{for } (x,y) \in L_1,
\end{aligned}
$$

$$
\begin{aligned}
\frac{\partial \phi}{\partial n} &= 0 \text{ for } (x,y) \in L_2, \\
\phi &= 0 \text{ for } (x,y) \in L_3 \cup L_4.
\end{aligned}
$$

It is easy to verify that the exact solution is given by

$$
\phi(x, y) = \sin(\pi y) \cosh(\pi[x - 1]).
$$

In view of the condition $\phi = 0$ for $(x,y) \in L_3 \cup L_4$, we may use the Green's function $\Phi_3(x, y; \xi, \eta)$ in Eq. (5.33) to obtain the boundary integral equation

$$
\begin{aligned}
\lambda(\xi, \eta)\phi(\xi, \eta) &= \int_{L_1 \cup L_2} [\phi(x,y)\frac{\partial}{\partial n}(\Phi_3(x, y; \xi, \eta)) \\
&\quad - \Phi_3(x, y; \xi, \eta)\frac{\partial}{\partial n}(\phi(x,y))]ds(x,y),
\end{aligned}
$$

with the parameter λ defined by

$$
\lambda(\xi, \eta) = \begin{cases} 0 & \text{if } (\xi, \eta) \in L_3 \cup L_4, \\ 1/2 & \text{if } (\xi, \eta) \text{ lies on a smooth part of } L_1 \cup L_2, \\ 1 & \text{if } (\xi, \eta) \in R. \end{cases}
$$

In using the above integral formulation to devise a boundary element procedure, we do not have to discretize L_3 and L_4. We discretize L_1 and L_2 into N straight line elements denoted by $C^{(1)}$, $C^{(2)}$, $\cdots$, $C^{(N-1)}$ and $C^{(N)}$. Proceeding as before, we may approximate ϕ and $\partial\phi/\partial n$ as constants over the boundary elements and reduce the task of finding the unknown ϕ on $L_1 \cup L_2$ to solving a system of linear algebraic equations. To set up the linear algebraic equations, we have to evaluate the line integrals

$$\int_{C^{(k)}} \Phi_3(x, y; \xi, \eta) ds(x, y) \text{ and } \int_{C^{(k)}} \frac{\partial}{\partial n}[\Phi_3(x, y; \xi, \eta)] ds(x, y).$$

These line integrals may be evaluated numerically using the formula in Eq. (3.23) (page 88, Chapter 3).

We shall not go into details here, but earlier FORTRAN codes such as CPG, CEHHZ1 and CEHHZ2 (listed on pages 91 to 95, Chapter 3) can be easily modified in an appropriate manner to solve the problem under consideration here through the special boundary integral formulation given above.

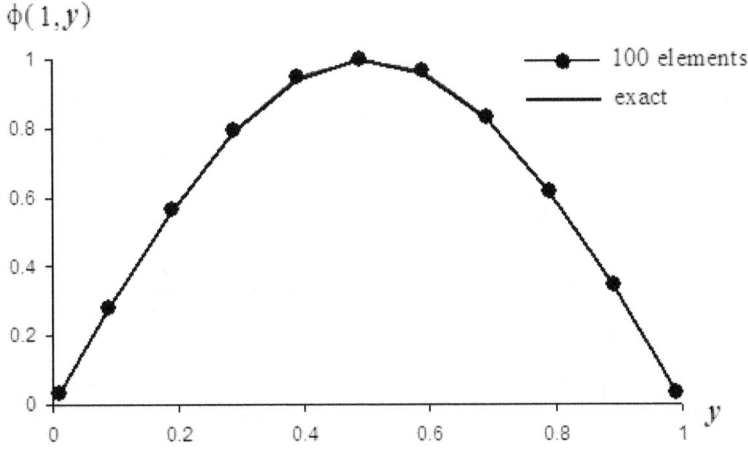

Figure 5.10

To obtain some numerical results, the semi-circle L_1 is discretized into 50 equal length boundary elements and the straight line L_2 into another 50 equal length elements (that is, 100 boundary elements are employed on $L_1 \cup L_2$). As $\partial\phi/\partial n$ is specified on L_2 ($x = 1$, $0 < y < 1$), we compare the numerically obtained values of $\phi(1, y)$ against the exact solution in Figure 5.10. There is a reasonably good agreement between the numerical and the exact values of ϕ on L_2. The validity of the Green's function $\Phi_3(x, y; \xi, \eta)$ is thus verified here.

Example 5.4

Take the solution domain R to be $0 < x < 3$, $0 < y < 1$. We are interested in solving Eq. (5.14) in R subject to the boundary conditions

$$\left.\frac{\partial\phi}{\partial n}\right|_{y=0} = 0 \text{ for } 0 < x < 3,$$

$$\left.\frac{\partial\phi}{\partial n}\right|_{y=1} = \begin{cases} 1 & \text{for } 1 < x < 2, \\ 0 & \text{otherwise,} \end{cases}$$

$$\phi(0, y) = 0 \text{ for } 0 < y < 1,$$

$$\left.\frac{\partial\phi}{\partial n}\right|_{x=3} = 0 \text{ for } 0 < y < 1.$$

If we use the Green's function $\Phi_4(x, y; \xi, \eta)$ in Eq. (5.35) to derive a boundary integral equation for the problem, we obtain

$$\lambda(\xi, \eta)\phi(\xi, \eta)$$

$$= -\int_1^2 \Phi_4(x, 1; \xi, \eta)dx - \int_0^1 \Phi_4(0, y; \xi, \eta)\left.\frac{\partial\phi}{\partial n}\right|_{x=0} dx$$

$$+ \int_0^1 \left.\frac{\partial}{\partial n}(\Phi_4(x, y; \xi, \eta))\right|_{x=3} \phi(3, y)dy, \qquad (5.37)$$

where $\lambda(\xi, \eta) = 1/2$ if (ξ, η) lies on the vertical sides ($x = 0$ and $x = 3$) of the rectangular solution domain and $\lambda(\xi, \eta) = 1$

186

if (ξ, η) lies on the horizontal sides ($y = 0$ and $y = 1$) or in the interior of the solution domain.

With the use of the special Green's function $\Phi_4(x, y; \xi, \eta)$, it is not necessary to integrate over the parts of the horizontal sides ($y = 0$ and $y = 1$) where $\partial\phi/\partial n$ is specified as 0. Proceeding as usual, we may discretize the boundary integral equation above to set up a system of linear algebraic equations to determine $\partial\phi/\partial n$ on $x = 0$ and ϕ on $x = 3$. We assume that $\partial\phi/\partial n$ is a constant over each element. Once the unknown values of $\partial\phi/\partial n$ on the elements are determined, $\phi(\xi, \eta)$ may be computed at any point (ξ, η) in the solution domain.

Table 5.1

(x, y)	With special Green's function	Without special Green's function
$(0.50, 0.25)$	0.4847	0.4858
$(1.50, 0.25)$	1.2666	1.2677
$(2.50, 0.25)$	1.4874	1.4831
$(0.50, 0.50)$	0.4978	0.4983
$(1.50, 0.50)$	1.3322	1.3340
$(2.50, 0.50)$	1.5015	1.4968
$(0.50, 0.75)$	0.5147	0.5129
$(1.50, 0.75)$	1.4556	1.4584
$(2.50, 0.75)$	1.5180	1.5127

It appears that there is no simple analytical solution for the problem under consideration. To check the validity of the special boundary integral equation obtained using the Green's function $\Phi_4(x, y; \xi, \eta)$, we compare its numerical values of ϕ with those obtained using the boundary element procedure outlined in Chapter 1 (that is, without the use of any special Green's function, by using the subroutines CELAP1 and CELAP2). The intervals of integration in the special boundary integral equation in Eq. (5.37) are discretized into 120 straight line elements, each of length 0.0250 units, to set up a system of 80

187

linear algebraic equations in 80 unknowns. For the boundary element method in Chapter 1, 320 elements, each also of length 0.0250 units, are employed. The numerical values of ϕ thus obtained at selected points in the interior of the solution domain are given in Table 5.1. There is a good agreement between the two sets of numerical values.

5.4 Exterior Region of a Circle

5.4.1 Two Special Green's Functions

Consider the region $x^2+y^2 > a^2$ (a is a positive real number) as sketched in Figure 5.11. For this region, we give Green's functions[5] satisfying certain conditions on the circular boundary $x^2 + y^2 = a^2$.

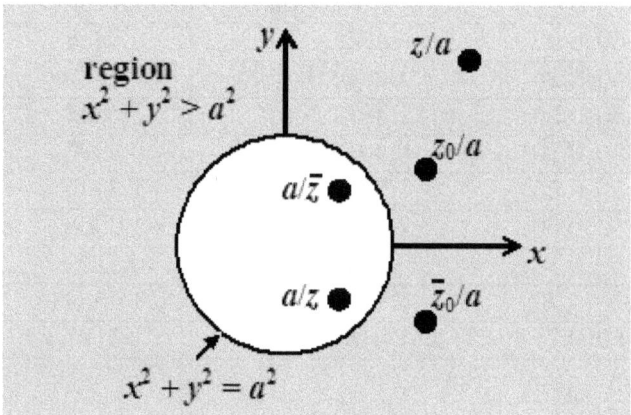

Figure 5.11

---

[5] *Note from the author.* I had no prior knowledge of the specific forms of the Green's functions $\Phi_5(x, y; \xi, \eta)$ and $\Phi_6(x, y; \xi, \eta)$ given in Eqs. (5.42) and (5.44). I deduced them by guesswork during the writing of this section. I believe that they must have already been recorded (in equivalent forms) somewhere in the research literature. If you have any information on this, please e-mail me at mwtang@ntu.edu.sg.

Let the required Green's functions take the form

$$\Phi(x, y; \xi, \eta)$$
$$= \frac{1}{4\pi} \ln([\frac{x}{a} - \frac{\xi}{a}]^2 + [\frac{y}{a} - \frac{\eta}{a}]^2) + \Phi^*(x, y; \xi, \eta), \quad (5.38)$$

such that $\Phi^*(x, y; \xi, \eta)$ satisfies

$$\frac{\partial^2}{\partial x^2}[\Phi^*(x, y; \xi, \eta)] + \frac{\partial^2}{\partial y^2}[\Phi^*(x, y; \xi, \eta)] = 0$$

$$\text{for } x^2 + y^2 > a^2 \text{ and } \xi^2 + \eta^2 > a^2.$$
$$(5.39)$$

For convenience, we define $z = x + iy$ and $z_0 = \xi + i\eta$ ($i = \sqrt{-1}$) and refer to the points (x, y) and (ξ, η) as z and z_0 respectively. We may rewrite $\Phi(x, y; \xi, \eta)$ in Eq. (5.38) given above as[6]

$$\Phi(x, y; \xi, \eta) = \frac{1}{2\pi} \operatorname{Re}\{\ln(\frac{1}{a}[z - z_0])\} + \Phi^*(x, y; \xi, \eta), \quad (5.40)$$

where Re denotes the real part of a complex number.

Let us take

$$\Phi^*(x, y; \xi, \eta) = -\frac{1}{2\pi} \operatorname{Re}\{\ln(1 - \frac{\overline{z_0}z}{a^2})\}, \quad (5.41)$$

where the overhead bar denotes the complex conjugate of a complex number.

From Figure 5.11, it is obvious that $a/z \neq \overline{z_0}/a$ for all points z and z_0 in the region $x^2 + y^2 > a^2$. This implies that $1 - \overline{z_0}z/a^2$ is not zero over the region $x^2 + y^2 > a^2$ and $\Phi^*(x, y; \xi, \eta)$ as given in Eq. (5.41) satisfies Eq. (5.39).

We may use Eqs. (5.40) and (5.41) to define a Green's function for the region $x^2 + y^2 > a^2$, that is,

$$\Phi_5(x, y; \xi, \eta)$$
$$= \frac{1}{2\pi} \operatorname{Re}\{\ln(\frac{1}{a}[z - z_0])\} - \frac{1}{2\pi} \operatorname{Re}\{\ln(1 - \frac{\overline{z_0}z}{a^2})\}.$$
$$(5.42)$$

---

[6]The complex logarithmic function $\ln(w)$ is defined by $\ln|w| + i \arg(w)$. Thus, $\operatorname{Re}\{\ln(z)\} = \frac{1}{2}\ln(p^2 + q^2)$ if $w = p + iq$, where p and q are real numbers.

If we write z in polar form as $z = r\exp(i\theta)$, we obtain

$$\Phi_5(x,y;\xi,\eta)|_{\text{on the circle } x^2+y^2=a^2}$$

$$= \frac{1}{2\pi}\operatorname{Re}\{\ln(\exp(i\theta) - \frac{z_0}{a})\}$$

$$-\frac{1}{2\pi}\operatorname{Re}\{\ln(1 - \frac{\overline{z}_0\exp(i\theta)}{a})\}$$

$$= \frac{1}{2\pi}\operatorname{Re}\{\ln(\exp(i\theta) - \frac{z_0}{a})\}$$

$$-\frac{1}{2\pi}\operatorname{Re}\{\overline{\ln(1 - \frac{\overline{z}_0\exp(i\theta)}{a})}\}$$

$$= \frac{1}{2\pi}\operatorname{Re}\{\ln(\exp(i\theta) - \frac{z_0}{a})\}$$

$$-\frac{1}{2\pi}\operatorname{Re}\{\ln(1 - \frac{z_0\exp(-i\theta)}{a})\}$$

$$= \frac{1}{2\pi}\operatorname{Re}\{\ln(\exp(i\theta) - \frac{z_0}{a})\}$$

$$-\frac{1}{2\pi}\operatorname{Re}\{\ln[\exp(-i\theta)(\exp(i\theta) - \frac{z_0}{a})]\}$$

$$= 0.$$

Note that $r = \sqrt{x^2 + y^2}$, that is, $r = a$ on the circle.

Thus, the Green's function $\Phi_5(x,y;\xi,\eta)$ in Eq. (5.42) satisfies the boundary condition

$$\Phi_5(x,y;\xi,\eta) = 0 \text{ on the circle } x^2 + y^2 = a^2. \qquad (5.43)$$

In a similar way, another Green's function for the region $x^2 + y^2 > a^2$ as defined by

$$\Phi_6(x,y;\xi,\eta) = \frac{1}{2\pi}\operatorname{Re}\{\ln(\frac{1}{a}[z - z_0])\}$$

$$+\frac{1}{2\pi}\operatorname{Re}\{\ln(1 - \frac{\overline{z}_0 z}{a^2}) - \ln(\frac{z}{a})\}$$

$$(5.44)$$

can be shown to satisfy the boundary condition

$$\frac{\partial}{\partial n}[\Phi_6(x,y;\xi,\eta)] = 0 \text{ on the circle } x^2 + y^2 = a^2. \qquad (5.45)$$

190

Note that $\partial[\Phi_6(x,y;\xi,\eta)]/\partial n = -\partial[\Phi_6(r\cos\theta, r\sin\theta;\xi,\eta)]/\partial r$ on $x^2 + y^2 = a^2$.

5.4.2 Applications

Example 5.5

The solution domain R is taken to be a doubly connected region given by

$$R = \{(x,y): \ x^2 + y^2 > 1, \ -\ell < x < \ell, \ -\ell < y < \ell, \ \ell > 1\}.$$

Refer to Figure 5.12.

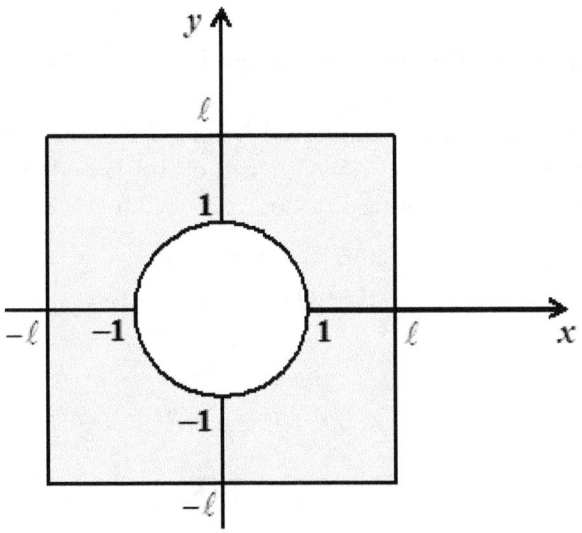

Figure 5.12

We are interested in solving Eq. (5.14) in R subject to the boundary conditions

$$\phi \ = \ 0 \text{ on the inner boundary } I \text{ (circle)},$$
$$\phi \ = \ 1 \text{ on the outer boundary } E \text{ (sides of square)}.$$

For the boundary value problem above, the Green's function $\Phi_5(x, y; \xi, \eta)$ in Eq. (5.42) may be applied together with the given boundary conditions to obtain the boundary integral equation

$$\lambda(\xi, \eta)\phi(\xi, \eta) = \int_E [\frac{\partial}{\partial n}(\Phi_5(x, y; \xi, \eta))$$

$$-\Phi_5(x, y; \xi, \eta)\frac{\partial}{\partial n}(\phi(x, y))]ds(x, y),$$

(5.46)

with the parameter λ defined by

$$\lambda(\xi, \eta) = \begin{cases} 0 & \text{if } (\xi, \eta) \in I, \\ 1/2 & \text{if } (\xi, \eta) \text{ lies on a smooth part of } E, \\ 1 & \text{if } (\xi, \eta) \in R. \end{cases}$$

Integration over the inner boundary I is avoided by the use of the special Green's function.

The problem has been solved before using several different approaches[7]. For $\ell = 1.53499$, a method based on complex variables gives the approximation[8]

$$\begin{aligned} \phi(x, y) &\simeq 0.99203 \cdot \ln(x^2 + y^2) \\ &+ 0.01331([x + iy]^4 + [x - iy]^4) \\ &\times (1 - \frac{1}{[x^2 + y^2]^4}) \\ &+ 0.00007([x + iy]^8 + [x - iy]^8) \\ &\times (1 - \frac{1}{[x^2 + y^2]^8}). \end{aligned}$$

(5.47)

---

[7]See, for example, the article "A solution of Laplace's equation for a round hole in a square peg," *Journal of the Society for Industrial and Applied Mathematics* (Volume 12, 1964, pp. 1-14) by RW Hockney. As pointed in this article, the boundary value problem under consideration arises in the modeling of gas leakage across the graphite brick of a nuclear reactor.

[8]This is as given in "Laplace's equation in the region bounded by a circle and a square," Technical Report No. M9/97 (Universiti Sains Malaysia, 1997) by KH Chew. Approximate formulae of ϕ for other values of ℓ may also found in this report.

We discretize the outer boundary E into 80 equal length elements to set up a system of 80 linear algebraic equations to determine $\partial\phi/\partial n$ on E. The unknown $\partial\phi/\partial n$ is assumed to be a constant over each boundary element. In Table 5.2, the numerical values of ϕ as obtained using Eq. (5.46) are compared with those computed from Eq. (5.47) at selected points in the interior of R. The two sets of numerical values of ϕ are in quite close agreement with each other.

Table 5.2

(x, y)	Boundary integral equation in Eq. (5.46)	Approximate formula in Eq. (5.47)
$(0.77781, 0.77781)$	0.16856	0.16854
$(-0.60000, 1.0392)$	0.34049	0.34027
$(0.22574, -1.2803)$	0.57292	0.57184
$(1.5000, 0.00000)$	0.94001	0.93756
$(0.00000, 1.4000)$	0.76681	0.76497
$(1.4142, 1.4142)$	0.97920	0.98683

Example 5.6

The solution domain R is as in Figure 5.12. As in Example 5.6, the inner boundary $x^2 + y^2 = 1$ is denoted by I and the outer boundary (the sides of the square) by E. The governing equation in R is the Laplace's equation in Eq. (5.14). We impose the condition $\partial\phi/\partial n = 0$ on the interior boundary I. At each and every point on the exterior boundary E, either ϕ or $\partial\phi/\partial n$ is suitably prescribed.

If the Green's function $\Phi_6(x, y; \xi, \eta)$ in Eq. (5.44) is applied together with the boundary condition on I to obtain the boundary integral equation for the problem under consideration, we obtain

$$\lambda(\xi,\eta)\phi(\xi,\eta) \;=\; \int_E [\phi(x,y)\frac{\partial}{\partial n}(\Phi_6(x,y;\xi,\eta))$$

$$-\Phi_6(x,y;\xi,\eta)\frac{\partial}{\partial n}(\phi(x,y))]ds(x,y),$$

$$(5.48)$$

with the parameter λ defined by

$$\lambda(\xi,\eta) = \begin{cases} 1/2 & \text{if } (\xi,\eta) \text{ lies on a smooth part of } E, \\ 1 & \text{if } (\xi,\eta) \in R \cup I. \end{cases}$$

For a particular solution of Eq. (5.14) satisfying the condition $\partial\phi/\partial n = 0$ on I, we take

$$\phi(x,y) = x + \frac{x}{x^2 + y^2}. \qquad (5.49)$$

We use Eq. (5.49) to produce boundary data for ϕ on E and discretize the boundary integral equation in Eq. (5.48) to solve numerically the problem under consideration subject to the boundary data of ϕ generated on E. If the numerical procecure really works, we should be able to recover numerically the solution in Eq. (5.49). For $\ell = 2.00000$, in Table 5.3, the numerical values of ϕ on I (where $r = 1$) obtained using 80 boundary elements are compared with the exact solution in Eq. (5.49) for selected values of the polar angle θ.

Table 5.3

θ	Eq. (5.48)	Exact solution
0°	2.000006	2.000000
15°	1.931910	1.931852
30°	1.732099	1.732051
45°	1.414239	1.414214
60°	1.000003	1.000000
75°	0.517632	0.517638
90°	0.000000	0.000000

5.5   Summary and Discussion

Several Green's functions for the two-dimensional Laplace's equation in a half plane, an infinitely long strip and a region exterior to a circle, which satisfy certain boundary conditions, are given. They are applied to obtain special boundary integral equations for particular boundary value problems.

As is clear from the examples given, if the required solution and the Green's function used satisfy the same homogeneous condition[9] on a certain part of the boundary, it is not necessary to integrate over that part. Furthermore, that boundary condition is automatically satisfied in the boundary integral formulation of the problem under consideration and requires no further treatment. This gives rise to a smaller system of linear algebraic equations in the boundary element procedure.

The smaller system definitely helps to ease the requirement on computer memory storage and precision. There are fewer coefficients to compute in setting up the system and less computer time is needed to invert a smaller matrix. More computer time is, however, required to compute *each* coefficient, as the Green's function assumes a form which is more complicated than the usual fundamental solution. In general, the integration of the Green's function and its normal derivative over an element has to be done numerically. Whether or not there is a significant overall reduction (or increase) in computer time needed to complete the boundary element procedure depends on how complicated the Green's function is.

5.6   Exercises

Unless otherwise stated, the two-dimensional Laplace's equation is the governing partial differential equation in all the exercises below.

1. Consider again the boundary value problem in Exercise 7

---

[9]In the examples given here, the homogeneous conditions involved are given by either $\phi = 0$ or $\partial\phi/\partial n = 0$ on the boundary. The boundary condition given by $\partial\phi/\partial n + k\phi = 0$ may also be considered.

of Chapter 1 (page 52). Use the Green's function $\Phi_1(x, y; \xi, \eta)$ in Eq.(5.9) to obtain a special boundary integral equation for the problem. Discretize the boundary integral equation in order to solve for ϕ numerically. Compare the numerical values of ϕ at selected points in the interior of the solution domain with the exact solution.

2. Let $z = x + iy$ and $w = u + iv$, where x, y, u and v are real variables. Check that the mapping $w = z^2$ transforms the quarter plane $x > 0$, $y > 0$ (on the Oxy plane) to the half plane $v > 0$ (on the Ouv plane). Find a Green's function $\Phi(x, y; \xi, \eta)$ for the quarter plane such that

$$\Phi(0, y; \xi, \eta) = 0 \text{ for } 0 < y < \infty,$$
$$\Phi(x, 0; \xi, \eta) = 0 \text{ for } 0 < x < \infty.$$

3. For the quarter plane in Exercise 2, construct a Green's function $\Phi(x, y; \xi, \eta)$ such that

$$\frac{\partial}{\partial x}[\Phi(x, y; \xi, \eta)]\Big|_{x=0} = 0 \text{ for } 0 < y < \infty,$$
$$\frac{\partial}{\partial y}[\Phi(x, y; \xi, \eta)]\Big|_{y=0} = 0 \text{ for } 0 < x < \infty.$$

4. The mapping $w = \cos(\pi z/a)$ transforms the half strip $0 < x < a$, $y > 0$ to the half plane $v > 0$. Find a Green's function $\Phi(x, y; \xi, \eta)$ for the half strip such that

$$\Phi(0, y; \xi, \eta) = 0 \text{ for } y > 0,$$
$$\Phi(a, y; \xi, \eta) = 0 \text{ for } y > 0,$$
$$\Phi(x, 0; \xi, \eta) = 0 \text{ for } 0 < x < a.$$

Use the Green's function to obtain a special boundary integral equation for the boundary value problem in Exercise 3 of Chapter 2 (page 78). Discretize the boundary integral equation in order to solve for ϕ numerically. Compare the numerical values of ϕ at selected points in the interior of the solution domain with the exact solution.

5. For the Helmholtz equation on page 82 in the half plane $y > 0$, construct a Green's function $\Omega(x, y; \xi, \eta)$ satisfying the boundary condition

$$\Omega(x, 0; \xi, \eta) = 0 \quad \text{for } 0 < x < \infty.$$

6. Repeat Exercise 5 with the boundary condition

$$\frac{\partial}{\partial y}[\Omega(x, y; \xi, \eta)]\bigg|_{y=0} = 0 \quad \text{for } 0 < x < \infty.$$

7. The Green's function $\Phi_5(x, y; \xi, \eta)$ in Eq. (5.42), but not $\Phi_6(x, y; \xi, \eta)$ in Eq. (5.44), is valid for the circle $x^2 + y^2 < a^2$. Explain why.

Chapter 6

Three-dimensional Problems

6.1   Introduction

The earlier chapters deal with only two-dimensional problems. In this chapter, we show how the analyses and boundary element procedures in Chapters 1 and 3 for the two-dimensional Laplace's and Helmholtz type equations can be extended to include three-dimensional problems. Depending on the geometry involved, the task of discretizing the surface boundary of a three-dimensional solution domain into elements may be a complicated one. The boundary is approximated using planar elements which are triangular in shape. Unknown functions on the boundary are approximated as constants over the elements. In general, the integration of the fundamental solution and its normal derivative over an element has to be carried out numerically.

6.2   Laplace's Equation

6.2.1   Boundary Value Problem

We are interested in solving the three-dimensional Laplace's equation

$$\frac{\partial^2 \phi}{\partial x^2} + \frac{\partial^2 \phi}{\partial y^2} + \frac{\partial^2 \phi}{\partial z^2} = 0 \text{ in } R, \qquad (6.1)$$

subject to the conditions

$$\begin{aligned} \phi &= f_1(x, y, z) \text{ for } (x, y, z) \in S_1, \\ \frac{\partial \phi}{\partial n} &= f_2(x, y, z) \text{ for } (x, y, z) \in S_2, \qquad (6.2) \end{aligned}$$

where R is the three-dimensional region bounded by the closed surface S, S_1 and S_2 are non-intersecting surfaces such that $S_1 \cup S_2 = S$ and f_1 and f_2 are suitably prescribed functions. Note that $\partial \phi / \partial n = n_x \partial \phi / \partial x + n_y \partial \phi / \partial y + n_z \partial \phi / \partial z$, where $[n_x, n_y, n_z]$ is the unit normal vector to the surface S pointing out of R.

6.2.2 Fundamental Solution

Using spherical coordinates r, θ and φ, centered about $(0, 0, 0)$, as determined by the relations $x = r \sin \varphi \cos \theta$, $y = r \sin \varphi \sin \theta$ and $z = r \cos \varphi$, we define

$$\psi(r, \theta, \varphi) = \phi(r \sin \varphi \cos \theta, r \sin \varphi \sin \theta, r \cos \varphi).$$

Note that $r = \sqrt{x^2 + y^2 + z^2}$. The Laplace's equation in Eq. (6.1) can be then written as

$$\frac{1}{r^2} \frac{\partial}{\partial r}\left(r^2 \frac{\partial \psi}{\partial r}\right) + \frac{1}{r^2 \sin^2 \varphi} \frac{\partial^2 \psi}{\partial \theta^2} + \frac{1}{r^2 \sin \varphi} \frac{\partial}{\partial \varphi}\left(\sin \varphi \frac{\partial \psi}{\partial \varphi}\right) = 0. \qquad (6.3)$$

For the case in which ψ is dependent only on the coordinate r, Eq. (6.3) reduces to the ordinary differential equation

$$\frac{d}{dr}\left(r^2 \frac{\partial \psi}{\partial r}\right) = 0. \qquad (6.4)$$

The general solution of Eq. (6.4) is

$$\psi = \frac{A}{r} + B, \tag{6.5}$$

where A and B are arbitrary constants. The solution is singular at the point $(x, y, z) = (0, 0, 0)$ (that is, at $r = 0$).

Shifting the center of the spherical coordinates in Eq. (6.5) to the general point (ξ, η, ζ), we choose a solution which is singular at $(x, y, z) = (\xi, \eta, \zeta)$, as given specifically by

$$\Phi_{3D}(x, y, z; \xi, \eta, \zeta) = -\frac{1}{4\pi\sqrt{(x - \xi)^2 + (y - \eta)^2 + (z - \zeta)^2}}, \tag{6.6}$$

to be the fundamental solution of Eq. (6.1).

6.2.3 Reciprocal Relation

If ϕ_1 and ϕ_2 are any two solutions of Eq. (6.1) in the region R bounded by the closed surface S, we can proceed as in Section 1.3 (page 11, Chapter 1) to derive the reciprocal relation

$$\iint\limits_{S} (\phi_2 \frac{\partial \phi_1}{\partial n} - \phi_1 \frac{\partial \phi_2}{\partial n}) ds(x, y, z) = 0, \tag{6.7}$$

where ds denotes the area of an infinitesimal element on S.

6.2.4 Boundary Integral Equation

The reciprocal relation in Eq. (6.7) may be applied to obtain a boundary integral equation for the boundary value problem here.

Guided by the analysis in Section 1.4 (page 12, Chapter 1), we take $\phi_1 = \Phi_{3D}(x, y, z; \xi, \eta, \zeta)$ (the fundamental solution in Eq. (6.6)) and $\phi_2 = \phi(x, y, z)$, where ϕ is the required solution of the interior boundary value problem defined by Eqs. (6.1)-(6.2). The reciprocal relation is only valid if (ξ, η, ζ) is not in

$R \cup S$. If (ξ, η, ζ) lies in the interior of R, the surface S in Eq. (6.7) can be replaced by $S \cup S_\varepsilon$, where S_ε is the spherical surface $(x - \xi)^2 + (y - \eta)^2 + (z - \zeta)^2 = \varepsilon^2$ (inside R), that is,

$$\iint\limits_{S \cup S_\varepsilon} (\phi_2 \frac{\partial \phi_1}{\partial n} - \phi_1 \frac{\partial \phi_2}{\partial n}) ds(x, y, z) = 0. \qquad (6.8)$$

Letting $\varepsilon \to 0^+$ in Eq. (6.8) gives

$$\iint\limits_{S} (\phi(x, y, z) \frac{\partial}{\partial n} [\Phi_{3D}(x, y, z; \xi, \eta, \zeta)]$$

$$- \Phi_{3D}(x, y, z; \xi, \eta, \zeta) \frac{\partial}{\partial n} [\phi(x, y, z)]) ds(x, y, z)$$

$$= - \lim_{\varepsilon \to 0^+} \iint\limits_{S_\varepsilon} (\phi(x, y, z) \frac{\partial}{\partial n} [\Phi_{3D}(x, y, z; \xi, \eta, \zeta)]$$

$$- \Phi_{3D}(x, y, z; \xi, \eta, \zeta) \frac{\partial}{\partial n} [\phi(x, y, z)]) ds(x, y, z). \quad (6.9)$$

To investigate the limit on the right hand side of Eq. (6.9), we expand $\phi(x, y, z)$ as a Taylor's series about (ξ, η, ζ), that is, we write

$$\phi(x, y, z) = \sum_{j=0}^{\infty} \sum_{m=0}^{j} \sum_{k=0}^{m} \frac{(x - \xi)^k (y - \eta)^{m-k} (z - \zeta)^{j-m}}{(j - m)! k! (m - k)!}$$

$$\times \frac{\partial^j \phi}{\partial x^k \partial y^{m-k} \partial z^{j-m}} \bigg|_{(x,y,z)=(\xi,\eta,\zeta)}. \qquad (6.10)$$

If we use the local spherical coordinates r, θ and φ, centered about (ξ, η, ζ), as defined by the relations $x - \xi = r \sin \varphi \cos \theta$, $y - \eta = r \sin \varphi \sin \theta$ and $z - \zeta = r \cos \varphi$, we find that Eq. (6.10)

$$\phi = \sum_{j=0}^{\infty} \sum_{m=0}^{j} \sum_{k=0}^{m} \frac{\varepsilon^j (\sin \varphi \cos \theta)^k (\sin \varphi \sin \theta)^{m-k} (\cos \varphi)^{j-m}}{(j - m)! k! (m - k)!}$$

$$\times \frac{\partial^j \phi}{\partial x^k \partial y^{m-k} \partial z^{j-m}} \bigg|_{(x,y,z)=(\xi,\eta,\zeta)} \quad \text{on } S_\varepsilon. \qquad (6.11)$$

Eq. (6.6) gives

$$\Phi_{3D} = -\frac{1}{4\pi\varepsilon} \text{ on } S_\varepsilon. \tag{6.12}$$

Since $\partial f/\partial n = -\partial f/\partial r$ on S_ε, we find that Eqs. (6.6) and (6.10) give

$$\frac{\partial \Phi_{3D}}{\partial n} = -\frac{1}{4\pi\varepsilon^2} \text{ on } S_\varepsilon, \tag{6.13}$$

and

$$
\begin{aligned}
\frac{\partial \phi}{\partial n} = & -\sum_{j=0}^{\infty}\sum_{m=0}^{j}\sum_{k=0}^{m}\frac{1}{(j-m)!k!(m-k)!} \\
& \times j\varepsilon^{j-1}(\sin\varphi\cos\theta)^k(\sin\varphi\sin\theta)^{m-k}(\cos\varphi)^{j-m} \\
& \times \left.\frac{\partial^j \phi}{\partial x^k \partial y^{m-k}\partial z^{j-m}}\right|_{(x,y,z)=(\xi,\eta,\zeta)} \quad \text{on } S_\varepsilon.
\end{aligned}
\tag{6.14}
$$

If we substitute Eqs. (6.11)-(6.14) into the limit on the right hand side of Eq. (6.9), we find that

$$
\begin{aligned}
& -\lim_{\varepsilon\to 0^+}\iint_{S_\varepsilon}(\phi(x,y,z)\frac{\partial}{\partial n}[\Phi_{3D}(x,y,z;\xi,\eta,\zeta)] \\
& \quad -\Phi_{3D}(x,y,z;\xi,\eta,\zeta)\frac{\partial}{\partial n}[\phi(x,y,z)])ds(x,y,z) \\
& = \lim_{\varepsilon\to 0^+}\frac{\phi(\xi,\eta,\zeta)}{4\pi\varepsilon^2}\iint_{S_\varepsilon}ds(x,y,z) \\
& = \lim_{\varepsilon\to 0^+}\frac{\phi(\xi,\eta,\zeta)}{4\pi\varepsilon^2}\times\text{(area of the spherical surface } S_\varepsilon) \\
& = \phi(\xi,\eta,\zeta) \text{ for } (\xi,\eta,\zeta)\in R.
\end{aligned}
$$

If (ξ,η,ζ) lies on a smooth part of S, we take S_ε to be the part of the spherical surface $(x-\xi)^2 + (y-\eta)^2 + (z-\zeta)^2 = \varepsilon^2$ inside R. As $\varepsilon \to 0^+$, we expect S_ε to tend to a hemisphere. If we proceed in a similar manner as outlined above to investigate the limit in Eq. (6.9), we obtain

$$
\begin{aligned}
- \lim_{\varepsilon \to 0^+} &\iint_{S_\varepsilon} (\phi(x,y,z) \frac{\partial}{\partial n} [\Phi_{3\mathrm{D}}(x,y,z;\xi,\eta,\zeta)] \\
&- \Phi_{3\mathrm{D}}(x,y,z;\xi,\eta,\zeta) \frac{\partial}{\partial n} [\phi(x,y,z)]) ds(x,y,z) \\
= \frac{1}{2} &\phi(\xi,\eta,\zeta) \text{ for } (\xi,\eta,\zeta) \text{ which lies on a smooth part of } S.
\end{aligned}
$$

Thus, for the problem under consideration, the boundary integral equation of interest is given by

$$
\begin{aligned}
\lambda(\xi,\eta,\zeta)&\phi(\xi,\eta,\zeta) \\
= \iint_S &(\phi(x,y,z) \frac{\partial}{\partial n} [\Phi_{3\mathrm{D}}(x,y,z;\xi,\eta,\zeta)] \\
&- \Phi_{3\mathrm{D}}(x,y,z;\xi,\eta,\zeta) \frac{\partial}{\partial n} [\phi(x,y,z)]) ds(x,y,z),
\end{aligned}
\tag{6.15}
$$

where

$$
\lambda(\xi,\eta) = \begin{cases} 0 & \text{if } (\xi,\eta) \notin R \cup S, \\ 1/2 & \text{if } (\xi,\eta) \text{ lies on a smooth part of } S, \\ 1 & \text{if } (\xi,\eta) \in R. \end{cases}
\tag{6.16}
$$

6.2.5 Boundary Element Method

The boundary S is discretized into N surface elements denoted by $S^{(1)}$, $S^{(2)}$, $\cdots$, $S^{(N-1)}$ and $S^{(N)}$. We approximate ϕ and $\partial\phi/\partial n$ as constants over each element, that is,

$$
\phi \simeq \overline{\phi}^{(k)} \text{ and } \frac{\partial \phi}{\partial n} \simeq \overline{p}^{(k)} \text{ for } (x,y,z) \in S^{(k)} \ (k = 1,2,\cdots,N).
\tag{6.17}
$$

Substituting Eq. (6.17) into Eq. (6.15), we obtain

$$\lambda(\xi,\eta,\zeta)\phi(\xi,\eta,\zeta) = \sum_{k=1}^{N}\{\overline{\phi}^{(k)}\mathcal{D}_2^{(k)}(\xi,\eta,\zeta) - \overline{p}_1^{(k)}\mathcal{D}_1^{(k)}(\xi,\eta,\zeta)\},$$

$$(6.18)$$

where

$$\mathcal{D}_1^{(k)}(\xi,\eta,\zeta) = \iint_{S^{(k)}} \Phi_{3D}(x,y,z;\xi,\eta,\zeta)ds(x,y,z),$$

$$\mathcal{D}_2^{(k)}(\xi,\eta,\zeta) = \iint_{S^{(k)}} \frac{\partial}{\partial n}[\Phi_{3D}(x,y,z;\xi,\eta,\zeta)]ds(x,y,z).$$

$$(6.19)$$

Let $(\overline{x}^{(m)},\overline{y}^{(m)},\overline{z}^{(m)})$ be a selected point in the interior of the element $S^{(m)}$. If we choose (ξ,η,ζ) in Eq. (6.18) to be given in turn by $(\overline{x}^{(1)},\overline{y}^{(1)},\overline{z}^{(1)})$, $(\overline{x}^{(2)},\overline{y}^{(2)},\overline{z}^{(2)})$, $\cdots$, $(\overline{x}^{(N-1)},\overline{y}^{(N-1)},\overline{z}^{(N-1)})$ and $(\overline{x}^{(N)},\overline{y}^{(N)},\overline{z}^{(N)})$, we obtain

$$\begin{aligned}\frac{1}{2}\overline{\phi}^{(m)} &= \sum_{k=1}^{N}\{\overline{\phi}^{(k)}\mathcal{D}_2^{(k)}(\overline{x}^{(m)},\overline{y}^{(m)},\overline{z}^{(m)}) \\ &\quad -\overline{p}^{(k)}\mathcal{D}_1^{(k)}(\overline{x}^{(m)},\overline{y}^{(m)},\overline{z}^{(m)})\} \\ &\quad \text{for } m = 1,2,\cdots,N.\end{aligned}$$

$$(6.20)$$

On a typical element $S^{(k)}$, either $\overline{\phi}^{(k)}$ or $\overline{p}^{(k)}$ is known. Thus, Eq. (6.20) constitutes a system of N linear algebraic equations containing the N unknowns. We may rewrite it as

$$\sum_{k=1}^{N}a^{(mk)}q^{(k)} = \sum_{k=1}^{N}b^{(mk)} \text{ for } m = 1,2,\cdots,N,$$

$$(6.21)$$

where $a^{(mk)}$, $b^{(mk)}$ and $q^{(k)}$ are defined by

$$
a^{(mk)} = \begin{cases} -\mathcal{D}_1^{(k)}(\overline{x}^{(m)}, \overline{y}^{(m)}, \overline{z}^{(m)}) \\ \qquad \text{if } \phi \text{ is specified over } S^{(k)}, \\ \mathcal{D}_2^{(k)}(\overline{x}^{(m)}, \overline{y}^{(m)}, \overline{z}^{(m)}) - \frac{1}{2}\delta^{(mk)} \\ \qquad \text{if } \partial\phi/\partial n \text{ is specified over } S^{(k)}, \end{cases}
$$

$$
b^{(mk)} = \begin{cases} \overline{\phi}^{(k)}(-\mathcal{D}_2^{(k)}(\overline{x}^{(m)}, \overline{y}^{(m)}, \overline{z}^{(m)}) + \frac{1}{2}\delta^{(mk)}) \\ \qquad \text{if } \phi \text{ is specified over } S^{(k)}, \\ \overline{p}^{(k)}\mathcal{D}_1^{(k)}(\overline{x}^{(m)}, \overline{y}^{(m)}, \overline{z}^{(m)}) \\ \qquad \text{if } \partial\phi/\partial n \text{ is specified over } S^{(k)}, \end{cases}
$$

$$
\delta^{(mk)} = \begin{cases} 0 & \text{if } m \neq k, \\ 1 & \text{if } m = k, \end{cases}
$$

$$
q^{(k)} = \begin{cases} \overline{p}^{(k)} & \text{if } \phi \text{ is specified over } S^{(k)}, \\ \overline{\phi}^{(k)} & \text{if } \partial\phi/\partial n \text{ is specified over } S^{(k)}. \end{cases} \tag{6.22}
$$

Note that $q^{(1)}, q^{(2)}, \cdots, q^{(N-1)}$ and $q^{(N)}$ are N unknowns to be determined. Once they are determined, ϕ can be computed at any point (ξ, η, ζ) by using Eq. (6.18) with $\lambda(\xi, \eta, \zeta) = 1$.

6.2.6 Computation of Integrals over Surface Elements

We take $S^{(k)}$ to be a planar element having the shape of a triangle. With reference to the coordinate system $Oxyz$, the vertices of $S^{(k)}$ are $(x_1^{(k)}, y_1^{(k)}, z_1^{(k)})$, $(x_2^{(k)}, y_2^{(k)}, z_2^{(k)})$ and $(x_3^{(k)}, y_3^{(k)}, z_3^{(k)})$, arranged in counter clockwise order, as seen by an observer looking at the surface S from outside R.

The components of the unit normal vector to $S^{(k)}$ pointing out of R are given by

$$
n_x^{(k)} = \frac{(y_2^{(k)} - y_1^{(k)})(z_3^{(k)} - z_1^{(k)}) - (z_2^{(k)} - z_1^{(k)})(y_3^{(k)} - y_1^{(k)})}{d^{(k)}}
$$

$$
n_y^{(k)} = \frac{(z_2^{(k)} - z_1^{(k)})(x_3^{(k)} - x_1^{(k)}) - (x_2^{(k)} - x_1^{(k)})(z_3^{(k)} - z_1^{(k)})}{d^{(k)}}
$$

$$
n_z^{(k)} = \frac{(x_2^{(k)} - x_1^{(k)})(y_3^{(k)} - y_1^{(k)}) - (y_2^{(k)} - y_1^{(k)})(x_3^{(k)} - x_1^{(k)})}{d^{(k)}},
$$

$$
\tag{6.23}
$$

where

$$
\begin{aligned}
d^{(k)} \;=\; & [((y_2^{(k)} - y_1^{(k)})(z_3^{(k)} - z_1^{(k)}) \\
& -(z_2^{(k)} - z_1^{(k)})(y_3^{(k)} - y_1^{(k)}))^2 \\
& +((z_2^{(k)} - z_1^{(k)})(x_3^{(k)} - x_1^{(k)}) \\
& -(x_2^{(k)} - x_1^{(k)})(z_3^{(k)} - z_1^{(k)}))^2 \\
& +((x_2^{(k)} - x_1^{(k)})(y_3^{(k)} - y_1^{(k)}) \\
& -(y_2^{(k)} - y_1^{(k)})(x_3^{(k)} - x_1^{(k)}))^2]^{1/2}.
\end{aligned} \tag{6.24}
$$

We parameterize points on $S^{(k)}$ using

$$
\begin{aligned}
(x, y, z) \;=\; & (X^{(k)}(u,v), Y^{(k)}(u,v), Z^{(k)}(u,v)) \\
& \text{for } 0 < u < 1 - v,\; 0 < v < 1, \tag{6.25}
\end{aligned}
$$

where

$$
\left.
\begin{aligned}
X^{(k)}(u,v) &= (x_2^{(k)} - x_1^{(k)})u + (x_3^{(k)} - x_1^{(k)})v + x_1^{(k)} \\
Y^{(k)}(u,v) &= (y_2^{(k)} - y_1^{(k)})u + (y_3^{(k)} - y_1^{(k)})v + y_1^{(k)} \\
Z^{(k)}(u,v) &= -(n_z^{(k)})^{-1}[n_x^{(k)}(X^{(k)}(u,v) - x_1^{(k)}) \\
&\quad + n_y^{(k)}(Y^{(k)}(u,v) - y_1^{(k)})] + z_1^{(k)}
\end{aligned}
\right\}
$$
$$
\text{if } |n_z^{(k)}| \geq 1/\sqrt{3},
$$

$$
\left.
\begin{aligned}
X^{(k)}(u,v) &= (x_2^{(k)} - x_1^{(k)})u + (x_3^{(k)} - x_1^{(k)})v + x_1^{(k)} \\
Z^{(k)}(u,v) &= (z_2^{(k)} - z_1^{(k)})u + (z_3^{(k)} - z_1^{(k)})v + z_1^{(k)} \\
Y^{(k)}(u,v) &= -(n_y^{(k)})^{-1}[n_x^{(k)}(X^{(k)}(u,v) - x_1^{(k)}) \\
&\quad + n_z^{(k)}(Z^{(k)}(u,v) - z_1^{(k)})] + y_1^{(k)}
\end{aligned}
\right\}
$$
$$
\text{if } |n_z^{(k)}| < 1/\sqrt{3} \text{ and } |n_y^{(k)}| \geq 1/\sqrt{3},
$$

$$
\left.
\begin{aligned}
Y^{(k)}(u,v) &= (y_2^{(k)} - y_1^{(k)})u + (y_3^{(k)} - y_1^{(k)})v + y_1^{(k)} \\
Z^{(k)}(u,v) &= (z_2^{(k)} - z_1^{(k)})u + (z_3^{(k)} - z_1^{(k)})v + z_1^{(k)} \\
X^{(k)}(u,v) &= -(n_x^{(k)})^{-1}[n_y^{(k)}(Y^{(k)}(u,v) - y_1^{(k)}) \\
&\quad + n_z^{(k)}(Z^{(k)}(u,v) - z_1^{(k)})] + x_1^{(k)}
\end{aligned}
\right\}
$$
$$
\text{if } |n_z^{(k)}| < 1/\sqrt{3} \text{ and } |n_y^{(k)}| < 1/\sqrt{3}. \tag{6.26}
$$

Since $(n_x^{(k)})^2 + (n_y^{(k)})^2 + (n_z^{(k)})^2 = 1$, we expect at least one of the three vector components $n_x^{(k)}$, $n_y^{(k)}$ and $n_z^{(k)}$ to have a magnitude which is greater than or equal to $1/\sqrt{3}$.

Transforming the integration over $S^{(k)}$ to the Ouv plane, we obtain

$$\mathcal{D}_1^{(k)}(\xi, \eta, \zeta) = \int_0^1 \int_0^{1-v} E_1(u, v) J^{(k)} du dv,$$

$$\mathcal{D}_2^{(k)}(\xi, \eta, \zeta) = \int_0^1 \int_0^{1-v} E_2(u, v) J^{(k)} du dv, \qquad (6.27)$$

where $E_1(u, v)$ and $E_2(u, v)$ are defined by

$$
\begin{aligned}
&E_1(u, v) \\
&= \Phi_{3D}(X^{(k)}(u, v), Y^{(k)}(u, v), Z^{(k)}(u, v); \xi, \eta, \zeta), \\
&E_2(u, v) \\
&= \frac{\partial}{\partial n}[\Phi_{3D}(x, y, z; \xi, \eta, \zeta)]\Big|_{(x,y,z)=(X^{(k)}(u,v),Y^{(k)}(u,v),Z^{(k)}(u,v))},
\end{aligned}
$$
$$(6.28)$$

and the Jacobian $J^{(k)}$ is a constant given by[1]

$$
\begin{aligned}
J^{(k)} &= 2\sqrt{\sigma^{(k)}(\sigma^{(k)} - \alpha^{(k)})(\sigma^{(k)} - \beta^{(k)})(\sigma^{(k)} - \gamma^{(k)})}, \\
\sigma^{(k)} &= \frac{\alpha^{(k)} + \beta^{(k)} + \gamma^{(k)}}{2}, \\
\alpha^{(k)} &= \sqrt{(x_1^{(k)} - x_2^{(k)})^2 + (y_1^{(k)} - y_2^{(k)})^2 + (z_1^{(k)} - z_2^{(k)})^2}, \\
\beta^{(k)} &= \sqrt{(x_2^{(k)} - x_3^{(k)})^2 + (y_2^{(k)} - y_3^{(k)})^2 + (z_2^{(k)} - z_3^{(k)})^2}, \\
\gamma^{(k)} &= \sqrt{(x_3^{(k)} - x_1^{(k)})^2 + (y_3^{(k)} - y_1^{(k)})^2 + (z_3^{(k)} - z_1^{(k)})^2}.
\end{aligned}
$$
$$(6.29)$$

Although it is possible to express the inner integrals in the double integrals in Eq. (6.27) in terms of standard elementary functions, we evaluate the double integrals directly by using a

---

[1]The Heron's formula which expresses the area of a triangle in terms of the lengths of the sides is used here to derive $J^{(k)}$.

numerical integration formula. We rewite Eq. (6.27) as

$$\mathcal{D}_1^{(k)}(\xi, \eta, \zeta) = \int\limits_0^1 \int\limits_0^1 E_1(t(1-v), v)(1-v)J^{(k)} dt dv,$$

$$\mathcal{D}_2^{(k)}(\xi, \eta, \zeta) = \int\limits_0^1 \int\limits_0^1 E_2(t(1-v), v)(1-v)J^{(k)} dt dv.$$

$$(6.30)$$

To evaluate the double integrals in Eq. (6.30) numerically, we use the Gaussian integration formula given by[2]

$$\int\limits_0^1 \int\limits_0^1 f(t, v) dt dv \simeq \frac{1}{16} \sum_{i=1}^{16} f(t_i, v_i), \qquad (6.31)$$

where

$$
\begin{aligned}
(t_1, v_1) &= (\frac{1}{4} + \frac{1}{4\sqrt{3}}, \frac{1}{4} + \frac{1}{4\sqrt{3}}), \\
(t_2, v_2) &= (\frac{1}{4} + \frac{1}{4\sqrt{3}}, \frac{1}{4} - \frac{1}{4\sqrt{3}}), \\
(t_3, v_3) &= (\frac{1}{4} - \frac{1}{4\sqrt{3}}, \frac{1}{4} + \frac{1}{4\sqrt{3}}), \\
(t_4, v_4) &= (\frac{1}{4} - \frac{1}{4\sqrt{3}}, \frac{1}{4} - \frac{1}{4\sqrt{3}}), \\
(t_5, v_5) &= (\frac{3}{4} + \frac{1}{4\sqrt{3}}, \frac{1}{4} + \frac{1}{4\sqrt{3}}), \\
(t_6, v_6) &= (\frac{3}{4} + \frac{1}{4\sqrt{3}}, \frac{1}{4} - \frac{1}{4\sqrt{3}}), \\
(t_7, v_7) &= (\frac{3}{4} - \frac{1}{4\sqrt{3}}, \frac{1}{4} + \frac{1}{4\sqrt{3}}), \\
(t_8, v_8) &= (\frac{3}{4} - \frac{1}{4\sqrt{3}}, \frac{1}{4} - \frac{1}{4\sqrt{3}}),
\end{aligned}
$$

[2]Refer to one of the integration formulae for square domain under section 25.4.62 in M Abramowitz and IA Stegun, *Handbook of Mathematical Functions* (Dover, 1970) to see how Eqs. (6.31) and (6.32) come about.

$$(t_9, v_9) = (\frac{3}{4} + \frac{1}{4\sqrt{3}}, \frac{3}{4} + \frac{1}{4\sqrt{3}}),$$

$$(t_{10}, v_{10}) = (\frac{3}{4} + \frac{1}{4\sqrt{3}}, \frac{3}{4} - \frac{1}{4\sqrt{3}}),$$

$$(t_{11}, v_{11}) = (\frac{3}{4} - \frac{1}{4\sqrt{3}}, \frac{3}{4} + \frac{1}{4\sqrt{3}}),$$

$$(t_{12}, v_{12}) = (\frac{3}{4} - \frac{1}{4\sqrt{3}}, \frac{3}{4} - \frac{1}{4\sqrt{3}}),$$

$$(t_{13}, v_{13}) = (\frac{1}{4} + \frac{1}{4\sqrt{3}}, \frac{3}{4} + \frac{1}{4\sqrt{3}}),$$

$$(t_{14}, v_{14}) = (\frac{1}{4} + \frac{1}{4\sqrt{3}}, \frac{3}{4} - \frac{1}{4\sqrt{3}}),$$

$$(t_{15}, v_{15}) = (\frac{1}{4} - \frac{1}{4\sqrt{3}}, \frac{3}{4} + \frac{1}{4\sqrt{3}}),$$

$$(t_{16}, v_{16}) = (\frac{1}{4} - \frac{1}{4\sqrt{3}}, \frac{3}{4} - \frac{1}{4\sqrt{3}}). \tag{6.32}$$

Recall that in order to set up the linear algebraic equations in Eq. (6.20) we are required to choose an interior point $(\overline{x}^{(m)}, \overline{y}^{(m)}, \overline{z}^{(m)})$ on the triangular element $S^{(m)}$. We choose

$$(\overline{x}^{(m)}, \overline{y}^{(m)}, \overline{z}^{(m)}) = (X^{(m)}(\frac{1}{4}, \frac{1}{2}), Y^{(m)}(\frac{1}{4}, \frac{1}{2}), Z^{(m)}(\frac{1}{4}, \frac{1}{2})). \tag{6.33}$$

Because of the choice in Eq. (6.33), if the point (ξ, η, ζ) in Eq.(6.30) is given by $(\overline{x}^{(k)}, \overline{y}^{(k)}, \overline{z}^{(k)})$, it is mapped to the midpoint $(t, v) = (1/2, 1/2)$ inside the integration domain $0 \le t \le 1$, $0 \le v \le 1$. The point $(t, v) = (1/2, 1/2)$ is where the integrands are singular if (ξ, η, ζ) in Eq.(6.30) is given by $(\overline{x}^{(k)}, \overline{y}^{(k)}, \overline{z}^{(k)})$. This does not pose any difficulty as none of the 16 points listed in Eq. (6.32) coincides with $(1/2, 1/2)$.

6.2.7 Implementation on Computer

The subroutine BE3D accepts the number of elements N and the three vertices of each of the N triangular elements as inputs. The value of N is stored in the integer variable N. The x, y and z

coordinates of the vertices are in the real arrays xt(1:N,1:3), yt(1:N,1:3) and zt(1:N,1:3) respectively. For example, the x, y and z coordinates of the second vertex of the 7th element are given by xt(7,2), yt(7,2) and zt(7,2) respectively. The first, second and third vertices of every element are arranged in a counter clockwise direction as seen by an observer outside the solution domain.

From the inputs, BE3D computes $n_x^{(k)}, n_y^{(k)}$ and $n_z^{(k)}$ in Eq. (6.23) and stores the values obtained in the real arrays nx(1:N), ny(1:N) and nz(1:N) respectively. It also calculates the Jacobian $J^{(k)}$ (in Eq. (6.29)) in the real array jac(1:N). The coefficients needed to evaluate the parametric functions $X^{(k)}(u,v)$, $Y^{(k)}(u,v)$ and $Z^{(k)}(u,v)$, as defined in Eq. (6.26), are set up in the real arrays pc(1:N,1:3,1:3). For example, pc(7,1,1), pc(7,2,1) and pc(7,3,1) store the values of the coefficients of u in $X^{(7)}(u,v)$, $Y^{(7)}(u,v)$ and $Z^{(7)}(u,v)$ respectively. Similarly, pc(7,1,2), pc(7,2,2) and pc(7,3,2) contain the values of the coefficients of v and pc(7,1,3), pc(7,2,3) and pc(7,3,3) the constant coefficients in the same three parametric functions. The collocation points $(\overline{x}^{(m)}, \overline{y}^{(m)}, \overline{z}^{(m)})$ needed for setting up Eqs. (6.21) are in the real arrays xm(1:N), ym(1:N) and zm(1:N). Lastly, the values of t_i and v_i listed in Eq. (6.32) are kept in the real arrays tg(1:16) and vg(1:16) respectively.

The arrays nx(1:N), ny(1:N), nz(1:N), xm(1:N), ym(1:N) and zm(1:N) are returned as outputs of the subroutine BE3D. The computed values in the arrays jac(1:N), pc(1:N,1:3,1:3), tg(1:16) and vg(1:16) are declared common so that they are accessible by the subroutine CPD3D where $\mathcal{D}_1^{(k)}(\xi, \eta, \zeta)$ and $\mathcal{D}_2^{(k)}(\xi, \eta, \zeta)$ in Eq. (6.30) are computed by the formula in Eq. (6.31).

For the execution of the boundary element procedure, BE3D has to be called once before the subroutines CE3D1 and CE3D2.

The subroutine BE3D is as listed below.

```
subroutine BE3D(N,xt,yt,zt,xm,ym,zm,nx,ny,nz)

integer N,i
```

```fortran
      double precision xt(1000,3),yt(1000,3),
     & zt(1000,3),nx(1000),ny(1000),nz(1000),jac(1000),
     & pc(1000,3,3),xm(1000),ym(1000),zm(1000),tg(16),
     & vg(16),x21,x31,y21,y31,z21,z31,aa,bb,cc,dd,s3

      common /elem3d/jac,pc,tg,vg

      s3=1d0/dsqrt(3d0)

      do 20 i=1,N

      x21=xt(i,2)-xt(i,1)
      x31=xt(i,3)-xt(i,1)
      y21=yt(i,2)-yt(i,1)
      y31=yt(i,3)-yt(i,1)
      z21=zt(i,2)-zt(i,1)
      z31=zt(i,3)-zt(i,1)
      nx(i)=y21*z31-z21*y31
      ny(i)=z21*x31-x21*z31
      nz(i)=x21*y31-y21*x31
      dd=dsqrt(nx(i)**2d0+ny(i)**2d0+nz(i)**2d0)
      nx(i)=nx(i)/dd
      ny(i)=ny(i)/dd
      nz(i)=nz(i)/dd

      aa=dsqrt(x21**2d0+y21**2d0+z21**2d0)
      bb=dsqrt((xt(i,2)-xt(i,3))**2d0
     & +(yt(i,2)-yt(i,3))**2d0
     & +(zt(i,2)-zt(i,3))**2d0)
      cc=dsqrt(x31**2d0+y31**2d0+z31**2d0)
      dd=0.5d0*(aa+bb+cc)
      jac(i)=2d0*dsqrt(dd*(dd-aa)*(dd-bb)*(dd-cc))

      if (dabs(nz(i)).ge.s3) then
      pc(i,1,1)=x21
      pc(i,1,2)=x31
```

```
   pc(i,1,3)=xt(i,1)
   pc(i,2,1)=y21
   pc(i,2,2)=y31
   pc(i,2,3)=yt(i,1)
   pc(i,3,1)=-(nx(i)*pc(i,1,1)
 & +ny(i)*pc(i,2,1))/nz(i)
   pc(i,3,2)=-(nx(i)*pc(i,1,2)
 & +ny(i)*pc(i,2,2))/nz(i)
   pc(i,3,3)=zt(i,1)
   else if (dabs(ny(i)).ge.s3) then
   pc(i,1,1)=x21
   pc(i,1,2)=x31
   pc(i,1,3)=xt(i,1)
   pc(i,3,1)=z21
   pc(i,3,2)=z31
   pc(i,3,3)=zt(i,1)
   pc(i,2,1)=-(nx(i)*pc(i,1,1)
 & +nz(i)*pc(i,3,1))/ny(i)
   pc(i,2,2)=-(nx(i)*pc(i,1,2)
 & +nz(i)*pc(i,3,2))/ny(i)
   pc(i,2,3)=yt(i,1)
   else
   pc(i,2,1)=y21
   pc(i,2,2)=y31
   pc(i,2,3)=yt(i,1)
   pc(i,3,1)=z21
   pc(i,3,2)=z31
   pc(i,3,3)=zt(i,1)
   pc(i,1,1)=-(nz(i)*pc(i,3,1)
 & +ny(i)*pc(i,2,1))/nx(i)
   pc(i,1,2)=-(nz(i)*pc(i,3,2)
 & +ny(i)*pc(i,2,2))/nx(i)
   pc(i,1,3)=xt(i,1)
   endif

   xm(i)=pc(i,1,1)*0.25d0+pc(i,1,2)*0.5d0
 & +pc(i,1,3)
```

```fortran
      ym(i)=pc(i,2,1)*0.25d0+pc(i,2,2)*0.5d0
     & +pc(i,2,3)
      zm(i)=pc(i,3,1)*0.25d0+pc(i,3,2)*0.5d0
     & +pc(i,3,3)

20 continue

      tg(1)=0.25d0*(1d0+s3)
      vg(1)=tg(1)
      tg(2)=tg(1)
      vg(2)=0.25d0*(1d0-s3)
      tg(3)=vg(2)
      vg(3)=tg(1)
      tg(4)=vg(2)
      vg(4)=vg(2)
      tg(5)=0.25d0*(3d0+s3)
      vg(5)=tg(1)
      tg(6)=tg(5)
      vg(6)=vg(2)
      tg(7)=0.25d0*(3d0-s3)
      vg(7)=tg(1)
      tg(8)=tg(7)
      vg(8)=vg(2)
      tg(9)=tg(5)
      vg(9)=tg(5)
      tg(10)=tg(5)
      vg(10)=tg(7)
      tg(11)=tg(7)
      vg(11)=tg(5)
      tg(12)=tg(7)
      vg(12)=tg(7)
      tg(13)=tg(1)
      vg(13)=tg(5)
      tg(14)=tg(1)
      vg(14)=tg(7)
      tg(15)=vg(2)
      vg(15)=tg(5)
```

```
tg(16)=vg(2)
vg(16)=tg(7)

return
end
```

The subroutine CPD3D computes $\mathcal{D}_1^{(k)}(\xi,\eta,\zeta)$ and $\mathcal{D}_2^{(k)}(\xi,\eta,\zeta)$ in Eq. (6.30) using the formula in Eq. (6.31). The values of ξ, η and ζ are inputs to the subroutine and are stored in the real variables xi, eta and zeta respectively. The integer k is also an input in the integer variable k. The real variables n1, n2 and n3 keep the values of the components of the normal vector of the k-th element. Thus, for example, if we use CPD3D to compute $\mathcal{D}_1^{(10)}(\xi,\eta,\zeta)$ and $\mathcal{D}_2^{(10)}(\xi,\eta,\zeta)$, n1, n2 and n3 as inputs contain the values of $n_x^{(10)}$, $n_y^{(10)}$ and $n_z^{(10)}$ respectively. The real variables D1 and D2 are outputs containing the values of $\mathcal{D}_1^{(k)}(\xi,\eta,\zeta)$ and $\mathcal{D}_2^{(k)}(\xi,\eta,\zeta)$ respectively.

The functions FS3D and DFS3D which compute $\Phi_{3D}(x, y, z; \xi, \eta, \zeta)$ (in Eq. (6.6)) and its normal derivative $\partial[\Phi_{3D}(x, y, z; \xi, \eta, \zeta)]/\partial n$ respectively are used in CPD3D.

The subroutine CPF3D and the functions FS3D and DFS3D are listed below.

```
subroutine CPD3D(xi,eta,zeta,k,n1,n2,n3,D1,D2)

integer k,N,i

double precision xi,eta,zeta,D1,D2,n1,n2,n3,
& jac(1000),pc(1000,3,3),tg(16),vg(16),
& FS3D,DFS3D,ut,xk,yk,zk

common /elem3d/jac,pc,tg,vg

D1=0d0
D2=0d0

do 10 i=1,16
```

```fortran
      ut=tg(i)*(1d0-vg(i))
      xk=pc(k,1,1)*ut+pc(k,1,2)*vg(i)+pc(k,1,3)
      yk=pc(k,2,1)*ut+pc(k,2,2)*vg(i)+pc(k,2,3)
      zk=pc(k,3,1)*ut+pc(k,3,2)*vg(i)+pc(k,3,3)
      D1=D1+FS3D(xk,yk,zk,xi,eta,zeta)*(1d0-vg(i))
      D2=D2+DFS3D(xk,yk,zk,xi,eta,zeta,n1,n2,n3)
     & *(1d0-vg(i))
10    continue

      D1=jac(k)*D1/16d0
      D2=jac(k)*D2/16d0

      return
      end

      function FS3D(x,y,z,xi,eta,ze)

      double precision FS3D,x,y,z,xi,eta,ze,pi

      pi=4d0*datan(1d0)
      FS3D=-0.25d0/
     & (pi*dsqrt((x-xi)**2d0+(y-eta)**2d0
     & +(z-ze)**2d0))

      return
      end

      function DFS3D(x,y,z,xi,eta,ze,nx,ny,nz)

      double precision DFS3D,x,y,z,xi,eta,ze,pi,
     & nx,ny,nz

      pi=4d0*datan(1d0)
      DFS3D=0.25d0*((x-xi)*nx+(y-eta)*ny+(z-ze)*nz)/
     & (pi*((x-xi)**2d0+(y-eta)**2d0
     & +(z-ze)**2d0)**1.5d0)
```

```
      return
      end
```

The boundary element procedure for finding unknown values of ϕ or $\partial\phi/\partial n$ on the elements are carried out in the subroutine CE3D1. The inputs required by CE3D1 are the integer variable N (number of elements) and the real arrays xm(1:N), ym(1:N) and zm(1:N) (coordinates of the collocation points $(\overline{x}^{(m)}, \overline{y}^{(m)}, \overline{z}^{(m)})$) and the real arrays nx(1:N), ny(1:N) and nz(1:N) (components of the normal vectors to the boundary elements). These real arrays are outputs of the subroutine BE3D. The integer arrays BCT(1:N) and the real arrays BCV(1:N) are also inputs to the subroutine CE3D1. As explained on page 39 in Chapter 1, the arrays BCT(1:N) and BCV(1:N) record the boundary conditions of the boundary value problem. Note that the subroutines CPD3D and SOLVER are called in CE3D1. The outputs of CE3D1 are the real arrays phi(1:N) and dphi(1:N) which contains respectively the values of ϕ and $\partial\phi/\partial n$ on the boundary elements.

```
      subroutine CE3D1(N,xm,ym,zm,nx,ny,nz,
     & BCT,BCV,phi,dphi)

      integer m,k,N,BCT(1000)

      double precision BCV(1000),A(1000,1000),
     & B(1000),D1,D2,del,phi(1000),dphi(1000),
     & Z(1000),nx(1000),ny(1000),nz(1000),
     & xm(1000),ym(1000),zm(1000)

      do 10 m=1,N
      B(m)=0d0
      do 5 k=1,N
      call CPD3D(xm(m),ym(m),zm(m),k,
     & nx(k),ny(k),nz(k),D1,D2)
      if (k.eq.m) then
      del=1d0
```

```
        else
        del=0d0
        endif
        if (BCT(k).eq.0) then
        A(m,k)=-D1
        B(m)=B(m)+BCV(k)*(-D2+0.5d0*del)
        else
        A(m,k)=D2-0.5d0*del
        B(m)=B(m)+BCV(k)*D1
        endif
5     continue
10    continue

        call solver(A,B,N,1,Z)

        do 15 m=1,N
        if (BCT(m).eq.0) then
        phi(m)=BCV(m)
        dphi(m)=Z(m)
        else
        phi(m)=Z(m)
        dphi(m)=BCV(m)
        endif
15    continue

        return
        end
```

After the execution of the boundary element procedure in
CE3D1, the subroutine CE3D2 may be called to compute numeri-
cally the solution ϕ at any interior point (ξ, η, ζ). The inputs required by CE3D2 are the integer variable N (number of elements),
the real variables xi, eta and zeta for ξ, η and ζ respectively,
the real arrays nx(1:N), ny(1:N) and nz(1:N) containing components of the normal vectors to the elements (returned by the
subroutine BE3D), and the real arrays phi(1:N) and dphi(1:N)

217

storing the values of ϕ and $\partial\phi/\partial n$ over the elements (as calculated by CE3D1). The numerical value of ϕ at (ξ, η, ζ) is returned by CE3D2 in the real variable `pint`.

```
    subroutine CE3D2(N,xi,eta,zeta,nx,ny,nz,
& phi,dphi,pint)

    integer N,i

    double precision xi,eta,zeta,
& nx(1000),ny(1000),nz(1000),
& phi(1000),dphi(1000),pint,sum,D1,D2

    sum=0d0

    do 10 i=1,N
    call CPD3D(xi,eta,zeta,i,nx(i),ny(i),nz(i),
& D1,D2)
    sum=sum+phi(i)*D2-dphi(i)*D1
10  continue

    pint=sum

    return
    end
```

Example 6.1

Take the solution domain to be given by $0 < x < 1$, $0 < y < 1$, $0 < z < 1$. For a specific test problem, we use the particular solution of Eq. (6.1) given by

$$\phi(x, y, z) = \Phi_{3D}\left(x, y, z; 0, 0, \frac{3}{2}\right),$$

to generate boundary data for ϕ on the face $x = 0$ ($0 < y < 1$, $0 < z < 1$) and $\partial\phi/\partial n$ on the remaining faces. (The boundary of the chosen solution domain comprises six faces.) Note that the particular solution is well defined at all points inside the solution domain. The boundary element procedure above is applied to solve Eq. (6.1) subject to the generated boundary data.

The faces of the cuboid region $0 < x < 1$, $0 < y < 1$, $0 < z < 1$ are discretized into $12N_0^2$ elements having the shape of isosceles right angle triangles. For $N_0 = 3$, the triangular elements (with their vertices) on one of the faces is as shown in Figure 6.1.

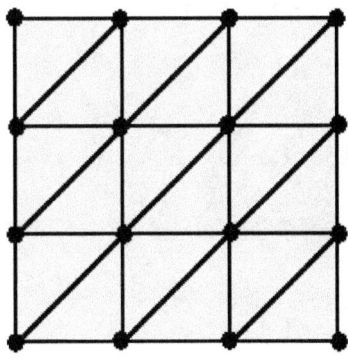

Figure 6.1

In the main program EX6PT1 listed below, the value of the positive integer N_0 is stored in the integer variable N0 and the coordinates of the three vertices of the triangular elements are in the real arrays xt(1:N,1:3), yt(1:N,1:3) and zt(1:N,1:3), as explained earlier on. A large part of the main program EX6PT1 is on the setting up of the arrays xt(1:N,1:3), yt(1:N,1:3) and zt(1:N,1:3) for the triangular elements on the six faces of the cuboid region $0 < x < 1$, $0 < y < 1$, $0 < z < 1$. To generate the required boundary data on the elements, we need to know the collocation points $(\overline{x}^{(m)}, \overline{y}^{(m)}, \overline{z}^{(m)})$ first. Thus, the subroutine BE3D is called before values are assigned to the real arrays BCV(1:N).

A complete listing of EX6PT1 is as follows.

```fortran
      program EX6PT1

      integer N0,BCT(1000),N,i,j,ians,ie

      double precision BCV(1000),phi(1000),
     & dphi(1000),pint,xi,
     & s1,s2,dl,nx(1000),ny(1000),nz(1000),eta,zeta,
     & xm(1000),ym(1000),zm(1000),
     & xt(1000,3),yt(1000,3),zt(1000,3),FS3D,DFS3D

      print*,'Enter N0 (<10):'
      read*,N0
      N=12*N0*N0

      dl=1d0/dfloat(2*N0)
      ie=-2

      do 10 i=1,N0
      do 10 j=1,N0
      ie=ie+2
      s1=dfloat(2*i-1)*dl
      s2=dfloat(2*j-1)*dl
      xt(ie+1,1)=0d0
      yt(ie+1,1)=s1-dl
      zt(ie+1,1)=s2-dl
      xt(ie+1,2)=0d0
      yt(ie+1,2)=s1+dl
      zt(ie+1,2)=s2+dl
      xt(ie+1,3)=0d0
      yt(ie+1,3)=s1+dl
      zt(ie+1,3)=s2-dl
      xt(ie+2,1)=0d0
      yt(ie+2,1)=s1-dl
      zt(ie+2,1)=s2-dl
      xt(ie+2,2)=0d0
      yt(ie+2,2)=s1-dl
      zt(ie+2,2)=s2+dl
```

```
xt(ie+2,3)=0d0
yt(ie+2,3)=s1+dl
zt(ie+2,3)=s2+dl
xt(2*N0*N0+ie+1,1)=1d0
yt(2*N0*N0+ie+1,1)=s1-dl
zt(2*N0*N0+ie+1,1)=s2-dl
xt(2*N0*N0+ie+1,2)=1d0
yt(2*N0*N0+ie+1,2)=s1+dl
zt(2*N0*N0+ie+1,2)=s2-dl
xt(2*N0*N0+ie+1,3)=1d0
yt(2*N0*N0+ie+1,3)=s1+dl
zt(2*N0*N0+ie+1,3)=s2+dl
xt(2*N0*N0+ie+2,1)=1d0
yt(2*N0*N0+ie+2,1)=s1-dl
zt(2*N0*N0+ie+2,1)=s2-dl
xt(2*N0*N0+ie+2,2)=1d0
yt(2*N0*N0+ie+2,2)=s1+dl
zt(2*N0*N0+ie+2,2)=s2+dl
xt(2*N0*N0+ie+2,3)=1d0
yt(2*N0*N0+ie+2,3)=s1-dl
zt(2*N0*N0+ie+2,3)=s2+dl
xt(4*N0*N0+ie+1,1)=s1+dl
yt(4*N0*N0+ie+1,1)=0d0
zt(4*N0*N0+ie+1,1)=s2-dl
xt(4*N0*N0+ie+1,2)=s1-dl
yt(4*N0*N0+ie+1,2)=0d0
zt(4*N0*N0+ie+1,2)=s2+dl
xt(4*N0*N0+ie+1,3)=s1-dl
yt(4*N0*N0+ie+1,3)=0d0
zt(4*N0*N0+ie+1,3)=s2-dl
xt(4*N0*N0+ie+2,1)=s1+dl
yt(4*N0*N0+ie+2,1)=0d0
zt(4*N0*N0+ie+2,1)=s2-dl
xt(4*N0*N0+ie+2,2)=s1+dl
yt(4*N0*N0+ie+2,2)=0d0
zt(4*N0*N0+ie+2,2)=s2+dl
xt(4*N0*N0+ie+2,3)=s1-dl
```

```
yt(4*N0*N0+ie+2,3)=0d0
zt(4*N0*N0+ie+2,3)=s2+dl
xt(6*N0*N0+ie+1,1)=s1-dl
yt(6*N0*N0+ie+1,1)=1d0
zt(6*N0*N0+ie+1,1)=s2-dl
xt(6*N0*N0+ie+1,2)=s1-dl
yt(6*N0*N0+ie+1,2)=1d0
zt(6*N0*N0+ie+1,2)=s2+dl
xt(6*N0*N0+ie+1,3)=s1+dl
yt(6*N0*N0+ie+1,3)=1d0
zt(6*N0*N0+ie+1,3)=s2-dl
xt(6*N0*N0+ie+2,1)=s1-dl
yt(6*N0*N0+ie+2,1)=1d0
zt(6*N0*N0+ie+2,1)=s2+dl
xt(6*N0*N0+ie+2,2)=s1+dl
yt(6*N0*N0+ie+2,2)=1d0
zt(6*N0*N0+ie+2,2)=s2+dl
xt(6*N0*N0+ie+2,3)=s1+dl
yt(6*N0*N0+ie+2,3)=1d0
zt(6*N0*N0+ie+2,3)=s2-dl
xt(8*N0*N0+ie+1,1)=s1-dl
yt(8*N0*N0+ie+1,1)=s2-dl
zt(8*N0*N0+ie+1,1)=0d0
xt(8*N0*N0+ie+1,2)=s1+dl
yt(8*N0*N0+ie+1,2)=s2+dl
zt(8*N0*N0+ie+1,2)=0d0
xt(8*N0*N0+ie+1,3)=s1+dl
yt(8*N0*N0+ie+1,3)=s2-dl
zt(8*N0*N0+ie+1,3)=0d0
xt(8*N0*N0+ie+2,1)=s1-dl
yt(8*N0*N0+ie+2,1)=s2-dl
zt(8*N0*N0+ie+2,1)=0d0
xt(8*N0*N0+ie+2,2)=s1-dl
yt(8*N0*N0+ie+2,2)=s2+dl
zt(8*N0*N0+ie+2,2)=0d0
xt(8*N0*N0+ie+2,3)=s1+dl
yt(8*N0*N0+ie+2,3)=s2+dl
```

```
      zt(8*N0*N0+ie+2,3)=0d0
      xt(10*N0*N0+ie+1,1)=s1-dl
      yt(10*N0*N0+ie+1,1)=s2-dl
      zt(10*N0*N0+ie+1,1)=1d0
      xt(10*N0*N0+ie+1,2)=s1+dl
      yt(10*N0*N0+ie+1,2)=s2-dl
      zt(10*N0*N0+ie+1,2)=1d0
      xt(10*N0*N0+ie+1,3)=s1+dl
      yt(10*N0*N0+ie+1,3)=s2+dl
      zt(10*N0*N0+ie+1,3)=1d0
      xt(10*N0*N0+ie+2,1)=s1-dl
      yt(10*N0*N0+ie+2,1)=s2-dl
      zt(10*N0*N0+ie+2,1)=1d0
      xt(10*N0*N0+ie+2,2)=s1+dl
      yt(10*N0*N0+ie+2,2)=s2+dl
      zt(10*N0*N0+ie+2,2)=1d0
      xt(10*N0*N0+ie+2,3)=s1-dl
      yt(10*N0*N0+ie+2,3)=s2+dl
      zt(10*N0*N0+ie+2,3)=1d0
10    continue

      call BE3D(N,xt,yt,zt,xm,ym,zm,nx,ny,nz)

      do 30 i=1,N
      if (i.le.(2*N0*N0)) then
      BCT(i)=0
      BCV(i)=FS3D(xm(i),ym(i),zm(i),0d0,0d0,1.5d0)
      else
      BCT(i)=1
      BCV(i)=DFS3D(xm(i),ym(i),zm(i),0d0,0d0,1.5d0,
   &  nx(i),ny(i),nz(i))
      endif
30    continue

      call CE3D1(N,xm,ym,zm,nx,ny,nz,BCT,BCV,
   &  phi,dphi)
```

```
50 print*,'Enter coordinates xi,eta and zeta of
&   an interior point:'

   read*,xi,eta,zeta

   call CE3D2(N,xi,eta,zeta,nx,ny,nz,
& phi,dphi,pint)

   write(*,60)pint,FS3D(xi,eta,zeta,0d0,0d0,1.5d0)

60 format('Numerical and exact values are:',
& F14.6,' and',F14.6,' respectively')

   print*,'To continue with another point enter 1:'
   read*,ians

   if (ians.eq.1) goto 50

   end
```

Table 6.1

(x, y, z)	$N_0 = 3$	$N_0 = 9$	Exact
$(0.50, 0.30, 0.10)$	-0.053551	-0.052588	-0.052472
$(0.20, 0.10, 0.70)$	-0.096167	-0.095623	-0.095800
$(0.10, 0.10, 0.90)$	-0.121527	-0.128574	-0.129092
$(0.10, 0.10, 0.95)$	-0.118659	-0.139119	-0.140128
$(0.05, 0.05, 0.95)$	-0.144826	-0.143637	-0.143505

The main program EX6PT1 is compiled together with the
subroutines BE3D, CE3D1, CE3D2, CPD3D and SOLVER (together
with its supporting subprograms) and the functions FS3D and
DFS3D into an executable code. The code is used to calculate
numerically ϕ in the interior of the cuboid region. In Table 6.1,
numerical values of ϕ obtained using $N_0 = 3$ and $N_0 = 9$ (that
is, using 108 and 972 triangular elements) are compared with
the exact solution $\phi = \Phi_{3D}(x, y, z; 0, 0, 3/2)$ at selected points

(x, y, z) in the interior of the solution domain. It is obvious that there is a significant accuracy in the numerical values as the number of boundary elements used is increased from 108 to 972. For points close to the boundary, it may be necessary to use more elements to obtain accurate numerical values of ϕ.

6.3 Homogeneous Helmholtz Equation

In place of the Laplace's equation in Eq. (6.1), let the boundary value problem in Section 6.2 be governed by the three-dimensional homogeneous Helmholtz equation

$$\frac{\partial^2 \phi}{\partial x^2} + \frac{\partial^2 \phi}{\partial y^2} + \frac{\partial^2 \phi}{\partial z^2} + w^2 \phi = 0, \tag{6.34}$$

where w is a real constant.

For the boundary value problem here, guided by the analysis in Section 3.1 (page 83, Chapter 3), one may proceed as above to derive the boundary integral equation

$$
\begin{aligned}
&\lambda(\xi, \eta, \zeta) \phi(\xi, \eta, \zeta) \\
&= \iint\limits_{S} (\phi(x, y, z) \frac{\partial}{\partial n} [\Omega_{3D}(x, y, z; \xi, \eta, \zeta)] \\
&\quad - \Omega_{3D}(x, y, z; \xi, \eta, \zeta) \frac{\partial}{\partial n} [\phi(x, y, z)]) ds(x, y, z),
\end{aligned}
\tag{6.35}
$$

where λ is as defined in Eq. (6.16) and $\Omega_{3D}(x, y, z; \xi, \eta, \zeta)$ is the fundamental solution of the three-dimensional Helmholtz equation as given by

$$\Omega_{3D}(x, y, z; \xi, \eta, \zeta) = -\frac{\cos(w \sqrt{(x - \xi)^2 + (y - \eta)^2 + (z - \zeta)^2})}{4\pi \sqrt{(x - \xi)^2 + (y - \eta)^2 + (z - \zeta)^2}}. \tag{6.36}$$

The integral equation in Eq. (6.35) may be discretized as in Section 6.2 to obtain a boundary element procedure for the numerical solution of Eq. (6.34). The general computer codes for the three-dimensional Laplace's equation may still be used here by modifying the functions FS3D and DFS3D to take into account the fundamental solution in Eq. (6.36)[3].

Example 6.2

The boundary integral equation in Eq. (6.35) is used here to solve numerically

$$\frac{\partial^2 \phi}{\partial x^2} + \frac{\partial^2 \phi}{\partial y^2} + \frac{\partial^2 \phi}{\partial z^2} + \frac{\pi^2}{16}\phi = 0$$
$$\text{in the region } x^2 + y^2 < 1,\ 0 < z < 1,$$

subject to

$$\left.\begin{array}{l}\phi(x,y,0) = 0 \\ \phi(x,y,1) = 1\end{array}\right\} \text{ for } x^2 + y^2 < 1,$$

$$\left.\frac{\partial \phi}{\partial n}\right|_{x^2+y^2=1} = 0 \text{ for } 0 < z < 1.$$

To discretize the boundary of the solution domain, define $\theta_m = 2(m-1)\pi/N_0$ $(m = 1, 2, \cdots, N_0)$ and $\ell_p = p/N_0$ $(p = 0, 1, 2, \cdots, N_0)$, where N_0 is an integer such that $N_0 \geq 3$. Put $N_0^2 + N_0$ points $(\cos\theta_n, \sin\theta_n, \ell_p)$ (for $n = 1, 2, \cdots, N_0$ and $p = 0, 1, 2, \cdots, N_0$) on the curved part of the boundary, and another $2N_0(N_0 - 1) + 2$ points $(0,0,0)$, $(0,0,1)$, $(\ell_k\cos\theta_n, \ell_k\sin\theta_n, 0)$ and $(\ell_k\cos\theta_n, \ell_k\sin\theta_n, 1)$ (for $n = 1, 2, \cdots, N_0$ and $k = 1, 2, \cdots, N_0 - 1$) on the flat parts of the boundary. (The curve part of the boundary is given by $x^2 + y^2 = 1$, $0 \leq z \leq 1$, while the flat parts by $x^2 + y^2 < 1$ with $z = 0$ and $z = 1$.)

On the curved part of the boundary, four typical neighboring points given by $(\cos\theta_{i-1}, \sin\theta_{i-1}, \ell_{k-1})$, $(\cos\theta_i, \sin\theta_i, \ell_{k-1})$, $(\cos\theta_{i-1}, \sin\theta_{i-1}, \ell_k)$ and $(\cos\theta_i, \sin\theta_i, \ell_k)$ are used to form 2 triangular elements as shown in Figure 6.2. This gives rise to $2N_0^2$ triangular elements.

---

[3]The numerical integration of the fundamental solution and its normal derivative may, however, have to be refined, if the magnitude of the coefficient w in Eq. (6.34) exceeds a certain value.

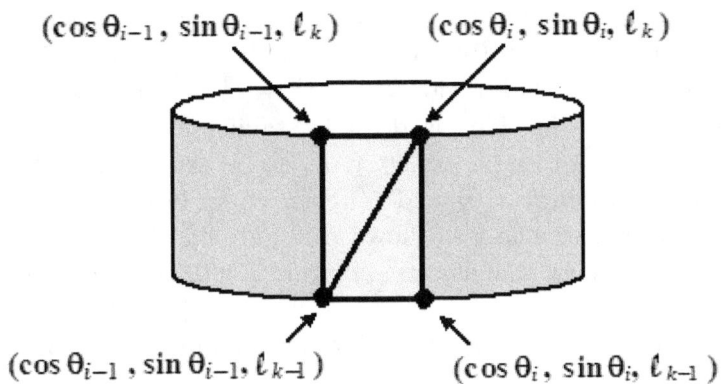

$(\cos\theta_{i-1}, \sin\theta_{i-1}, \ell_k)$ $(\cos\theta_i, \sin\theta_i, \ell_k)$

$(\cos\theta_{i-1}, \sin\theta_{i-1}, \ell_{k-1})$ $(\cos\theta_i, \sin\theta_i, \ell_{k-1})$

Figure 6.2

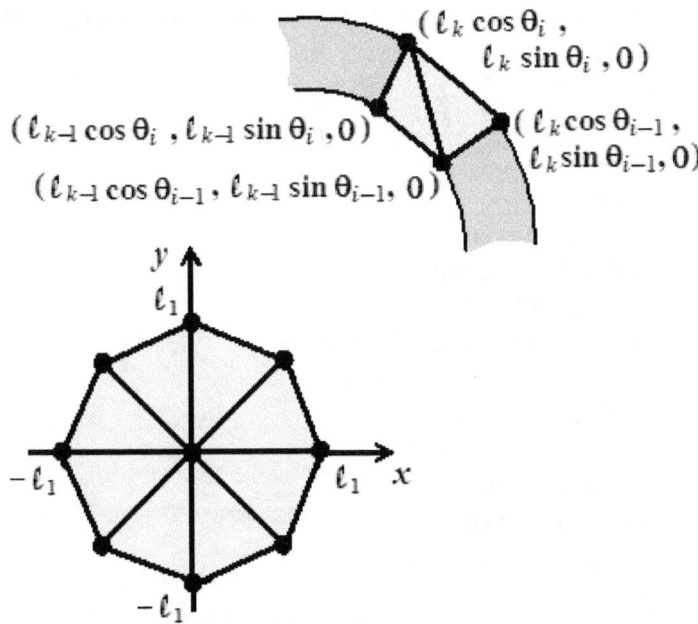

$(\ell_k \cos\theta_i, \ell_k \sin\theta_i, 0)$

$(\ell_{k-1}\cos\theta_i, \ell_{k-1}\sin\theta_i, 0)$ $(\ell_k\cos\theta_{i-1}, \ell_k\sin\theta_{i-1}, 0)$

$(\ell_{k-1}\cos\theta_{i-1}, \ell_{k-1}\sin\theta_{i-1}, 0)$

Figure 6.3

The flat part given by $x^2 + y^2 < 1$, $z = 0$ is discretized as follows. The region where $x^2 + y^2 \leq \ell_1$ is divided into N_0 triangular elements. All these elements share a common vertex given by $(0, 0, 0)$. A sketch of this is given in Figure 6.3 for $N_0 = 8$. The region $\ell_1 \leq x^2 + y^2 \leq 1$ is discretized into $2N_0(N_0 - 1)$. This is done by taking four neighboring points $(\ell_{k-1} \cos\theta_{i-1}, \ell_{k-1} \sin\theta_{i-1}, 0)$, $(\ell_{k-1} \cos\theta_i, \ell_{k-1} \sin\theta_i, 0)$, $(\ell_k \cos\theta_{i-1}, \ell_k \sin\theta_{i-1}, 0)$ and $(\ell_k \cos\theta_i, \ell_k \sin\theta_i, 0)$ $(2 \leq k \leq N_0)$ to form 2 elements as shown in Figure 6.3. The other flat part of boundary, that is, $x^2 + y^2 < 1$, $z = 0$, is discretized in a similar manner.

The total number of triangular elements involved is $6N_0^2 - 2N_0$.

The computer program for Example 6.1 can be easily modified for the problem here. The lines in the main program **EX6PT1** for setting up the vertices of the triangular elements and the boundary conditions are replaced by the codes listed below. Furthermore, the functions **FS3D** and **DFS3D** are modified to take into account the fundamental solution in Eq. (6.36) with $w = \pi/4$.

```
dl=1d0/dfloat(N0)
ie=-2
ig=-2
pi=4d0*datan(1d0)
N=6*N0*N0-2*N0
do 10 i=1,N0
s1=dfloat(2*(i-1))*pi*dl
s2=dfloat(2*i)*pi*dl
xt(i,1)=dl*dcos(s1)
yt(i,1)=dl*dsin(s1)
zt(i,1)=1d0
xt(i,2)=dl*dcos(s2)
yt(i,2)=dl*dsin(s2)
zt(i,2)=1d0
xt(i,3)=0d0
yt(i,3)=0d0
```

```
zt(i,3)=1d0
BCT(i)=0
BCV(i)=1d0
xt(N0+i,1)=dl*dcos(s2)
yt(N0+i,1)=dl*dsin(s2)
zt(N0+i,1)=0d0
xt(N0+i,2)=dl*dcos(s1)
yt(N0+i,2)=dl*dsin(s1)
zt(N0+i,2)=0d0
xt(N0+i,3)=0d0
yt(N0+i,3)=0d0
zt(N0+i,3)=0d0
BCT(N0+i)=0
BCV(N0+i)=0d0
do 10 j=0,N0-1
ie=ie+2
v1=dfloat(j)*dl
v2=dfloat(j+1)*dl
if (j.ne.0) then
ig=ig+2
v1=dfloat(j)*dl
v2=dfloat(j+1)*dl
xt(2*N0+ig+1,1)=v1*dcos(s1)
yt(2*N0+ig+1,1)=v1*dsin(s1)
zt(2*N0+ig+1,1)=1d0
xt(2*N0+ig+1,2)=v2*dcos(s1)
yt(2*N0+ig+1,2)=v2*dsin(s1)
zt(2*N0+ig+1,2)=1d0
xt(2*N0+ig+1,3)=v2*dcos(s2)
yt(2*N0+ig+1,3)=v2*dsin(s2)
zt(2*N0+ig+1,3)=1d0
BCT(2*N0+ig+1)=0
BCV(2*N0+ig+1)=1d0
xt(2*N0+ig+2,1)=v1*dcos(s1)
yt(2*N0+ig+2,1)=v1*dsin(s1)
zt(2*N0+ig+2,1)=1d0
xt(2*N0+ig+2,2)=v2*dcos(s2)
```

```
yt(2*N0+ig+2,2)=v2*dsin(s2)
zt(2*N0+ig+2,2)=1d0
xt(2*N0+ig+2,3)=v1*dcos(s2)
yt(2*N0+ig+2,3)=v1*dsin(s2)
zt(2*N0+ig+2,3)=1d0
BCT(2*N0+ig+2)=0
BCV(2*N0+ig+2)=1d0
xt(2*N0*N0+ig+1,2)=v1*dcos(s1)
yt(2*N0*N0+ig+1,2)=v1*dsin(s1)
zt(2*N0*N0+ig+1,2)=0d0
xt(2*N0*N0+ig+1,1)=v2*dcos(s1)
yt(2*N0*N0+ig+1,1)=v2*dsin(s1)
zt(2*N0*N0+ig+1,1)=0d0
xt(2*N0*N0+ig+1,3)=v2*dcos(s2)
yt(2*N0*N0+ig+1,3)=v2*dsin(s2)
zt(2*N0*N0+ig+1,3)=0d0
BCT(2*N0*N0+ig+1)=0
BCV(2*N0*N0+ig+1)=0d0
xt(2*N0*N0+ig+2,2)=v1*dcos(s1)
yt(2*N0*N0+ig+2,2)=v1*dsin(s1)
zt(2*N0*N0+ig+2,2)=0d0
xt(2*N0*N0+ig+2,1)=v2*dcos(s2)
yt(2*N0*N0+ig+2,1)=v2*dsin(s2)
zt(2*N0*N0+ig+2,1)=0d0
xt(2*N0*N0+ig+2,3)=v1*dcos(s2)
yt(2*N0*N0+ig+2,3)=v1*dsin(s2)
zt(2*N0*N0+ig+2,3)=0d0
BCT(2*N0*N0+ig+2)=0
BCV(2*N0*N0+ig+2)=0d0
endif
xt(4*N0*N0-2*N0+ie+1,1)=dcos(s1)
yt(4*N0*N0-2*N0+ie+1,1)=dsin(s1)
zt(4*N0*N0-2*N0+ie+1,1)=v1
xt(4*N0*N0-2*N0+ie+1,2)=dcos(s2)
yt(4*N0*N0-2*N0+ie+1,2)=dsin(s2)
zt(4*N0*N0-2*N0+ie+1,2)=v1
xt(4*N0*N0-2*N0+ie+1,3)=dcos(s2)
```

```
yt(4*N0*N0-2*N0+ie+1,3)=dsin(s2)
zt(4*N0*N0-2*N0+ie+1,3)=v2
BCT(4*N0*N0-2*N0+ie+1)=1
BCV(4*N0*N0-2*N0+ie+1)=0d0
xt(4*N0*N0-2*N0+ie+2,1)=dcos(s1)
yt(4*N0*N0-2*N0+ie+2,1)=dsin(s1)
zt(4*N0*N0-2*N0+ie+2,1)=v1
xt(4*N0*N0-2*N0+ie+2,2)=dcos(s2)
yt(4*N0*N0-2*N0+ie+2,2)=dsin(s2)
zt(4*N0*N0-2*N0+ie+2,2)=v2
xt(4*N0*N0-2*N0+ie+2,3)=dcos(s1)
yt(4*N0*N0-2*N0+ie+2,3)=dsin(s1)
zt(4*N0*N0-2*N0+ie+2,3)=v2
BCT(4*N0*N0-2*N0+ie+2)=1
BCV(4*N0*N0-2*N0+ie+2)=0d0
10 continue
```

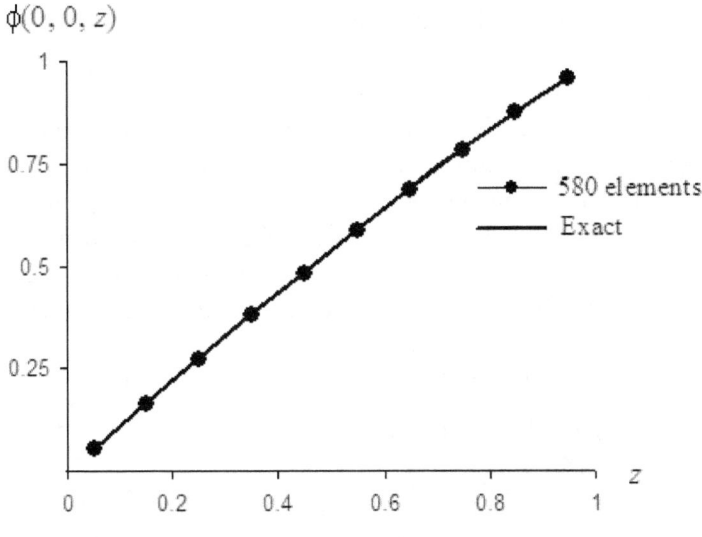

Figure 6.4

Numerical values of $\phi(0,0,z)$ obtained using $N_0 = 10$ (580 elements) are plotted against z for $0.0500 \leq z \leq 0.950$ and compared with the exact solution in Figure 6.4. The exact solution of the problem is given by $\phi(x, y, z) = \sqrt{2}\sin(\pi z/4)$. For $0.0500 \le z \le 0.950$, the numerical values agree with the exact ones to at least two decimal places. Thus, the two graphs in Figure 6.4 are visually indistinguishable.

6.4 Helmholtz Type Equation with Variable Coefficients

6.4.1 Dual-reciprocity Boundary Element Procedure

Let us consider now solving the generalized Helmholtz equation

$$\frac{\partial^2 \phi}{\partial x^2} + \frac{\partial^2 \phi}{\partial y^2} + \frac{\partial^2 \phi}{\partial z^2} + \alpha(x, y, z)\phi = g(x, y, z) \text{ in } R, \qquad (6.37)$$

subject to the boundary conditions in Eq. (6.2). As before, R denotes the three-dimensional region bounded by a closed surface S.

We show here how the dual-reciprocity boundary element method can be extended to solve the three-dimensional boundary value problem defined by Eqs. (6.37) and (6.2). The steps involved in the extension follow very closely the analysis given in Section 3.3 (page 99, Chapter 3). Thus, we shall not explain in great details how the main equations given here are obtained.

Firstly, Eq. (6.37) is used to derive the integral equation

$$
\begin{aligned}
\lambda(\xi, \eta, \zeta)&\phi(\xi, \eta, \zeta) \\
= \iiint_R &\Phi_{3D}(x, y, z; \xi, \eta, \zeta) \\
\times &[g(x, y, z) - \alpha(x, y, z)\phi(x, y, z)]dxdydz \\
+ \iint_S &[\phi(x, y, z)\frac{\partial}{\partial n}(\Phi_{3D}(x, y, z; \xi, \eta, \zeta)) \\
-&\Phi_{3D}(x, y, z; \xi, \eta, \zeta)\frac{\partial}{\partial n}(\phi(x, y, z))]ds(x, y, z) \\
&\text{for } (\xi, \eta, \zeta) \in R \cup S, \qquad (6.38)
\end{aligned}
$$

where $\Phi_{3D}(x, y, z; \xi, \eta, \zeta)$, as defined in Eq. (6.6), is the fundamental solution of the three-dimensional Laplace's equation.

Secondly, for treating the volume integral in Eq. (6.38), we use the radial basis function

$$\rho_{3D}(x, y, z; a, b, c) = 1 + r^2(x, y, z; a, b, c) + r^3(x, y, z; a, b, c),$$
(6.39)

and the corresponding function

$$\chi_{3D}(x, y, z; a, b, c) = \frac{1}{6}r^2(x, y, z; a, b, c) + \frac{1}{20}r^4(x, y, z; a, b, c)$$
$$+ \frac{1}{30}r^5(x, y, z; a, b, c),$$
(6.40)

where $r(x, y, z; a, b, c)$ denotes the distance between the point (x, y, z) and (a, b, c).

Lastly, as in Section 6.2, we discretize the surface boundary S into N triangular elements $S^{(1)}$, $S^{(2)}$, $\cdots$, $S^{(N-1)}$ and $S^{(N)}$ and approximate ϕ and $\partial\phi/\partial n$ as constants $\overline{\phi}^{(k)}$ and $\overline{p}_k$ respectively on the element $S^{(k)}$. For the dual-reciprocity method, the first N collocation points $(\overline{x}^{(1)}, \overline{y}^{(1)}, \overline{z}^{(1)})$, $(\overline{x}^{(2)}, \overline{y}^{(2)}, \overline{z}^{(2)})$, $\cdots$, $(\overline{x}^{(N-1)}, \overline{y}^{(N-1)}, \overline{z}^{(N-1)})$ and $(\overline{x}^{(N)}, \overline{y}^{(N)}, \overline{z}^{(N)})$ lie on the elements $S^{(1)}$, $S^{(2)}$, $\cdots$, $S^{(N-1)}$ and $S^{(N)}$ respectively. Another L collocation points denoted by $(\overline{x}^{(N+1)}, \overline{y}^{(N+1)}, \overline{z}^{(N+1)})$, $(\overline{x}^{(N+2)}, \overline{y}^{(N+2)}, \overline{z}^{(N+2)})$, $\cdots$, $(\overline{x}^{(N+L-1)}, \overline{y}^{(N+L-1)}, \overline{z}^{(N+L-1)})$ and $(\overline{x}^{(N+L)}, \overline{y}^{(N+L)}, \overline{z}^{(N+L)})$ are chosen from the interior of the solution domain R.

Eq. (6.38) can then be discretized to give the approximation

$$\lambda(\overline{x}^{(n)}, \overline{y}^{(n)}, \overline{z}^{(n)})\overline{\phi}^{(n)}$$

$$= \sum_{j=1}^{N+L} \mu_{3D}^{(nj)}[g(\overline{x}^{(j)}, \overline{y}^{(j)}, \overline{z}^{(j)}) - \alpha(\overline{x}^{(j)}, \overline{y}^{(j)}, \overline{z}^{(j)})\overline{\phi}^{(j)}]$$

$$+ \sum_{k=1}^{N} \{\overline{\phi}^{(k)}\mathcal{D}_2^{(k)}(\overline{x}^{(n)}, \overline{y}^{(n)}, \overline{z}^{(n)})$$

$$- \overline{p}^{(k)}\mathcal{D}_1^{(k)}(\overline{x}^{(n)}, \overline{y}^{(n)}, \overline{z}^{(n)})\}$$

$$\text{for } n = 1, 2, \cdots, N + L,$$
(6.41)

where $\mathcal{D}_1^{(k)}(\xi,\eta,\zeta)$ and $\mathcal{D}_2^{(k)}(\xi,\eta,\zeta)$ are as defined in Eq. (6.30) and $\mu_{3D}^{(nj)}$ are given by

$$\mu_{3D}^{(nj)} = \sum_{m=1}^{N+L} \omega_{3D}^{(mj)} \Psi_{3D}(\overline{x}^{(n)},\overline{y}^{(n)},\overline{z}^{(n)};\overline{x}^{(m)},\overline{y}^{(m)},\overline{z}^{(m)}), \quad (6.42)$$

together with

$$\sum_{j=1}^{N+L} \omega_{3D}^{(kj)} \rho_{3D}(\overline{x}^{(j)},\overline{y}^{(j)},\overline{z}^{(j)};\overline{x}^{(m)},\overline{y}^{(m)},\overline{z}^{(m)}) = \begin{cases} 1 & \text{if } k=m, \\ 0 & \text{if } k \neq m, \end{cases}$$

$$(6.43)$$

and

$$
\begin{aligned}
&\Psi_{3D}(\xi,\eta,\zeta;a,b,c) \\
=\ & \lambda(\xi,\eta,\zeta)\chi_{3D}(\xi,\eta,\zeta;a,b,c) \\
&+ \sum_{k=1}^{N}[n_x^{(k)}\frac{\partial}{\partial x}(\chi_{3D}(x,y,z;a,b,c)) \\
&+ n_y^{(k)}\frac{\partial}{\partial y}(\chi_{3D}(x,y,z;a,b,c)) \\
&+ \left. n_z^{(k)}\frac{\partial}{\partial z}(\chi_{3D}(x,y,z;a,b,c))]\right|_{(x,y,z)=(\overline{x}^{(k)},\overline{y}^{(k)},\overline{z}^{(k)})} \\
&\times \mathcal{D}_1^{(k)}(\xi,\eta,\zeta) \\
&- \sum_{k=1}^{N}\chi_{3D}(\overline{x}^{(k)},\overline{y}^{(k)},\overline{z}^{(k)};a,b,c)\mathcal{D}_2^{(k)}(\xi,\eta,\zeta). \quad (6.44)
\end{aligned}
$$

We rewrite Eq. (6.41) as

$$
\begin{aligned}
&\sum_{k=1}^{N+L} a^{(nk)}q^{(k)} \\
=\ & -\sum_{j=1}^{N+L} \mu_{3D}^{(nj)} g(\overline{x}^{(j)},\overline{y}^{(j)},\overline{z}^{(j)}) + \sum_{k=1}^{N} b^{(nk)} \\
&\text{for } n=1,2,\cdots,N+L, \quad (6.45)
\end{aligned}
$$

where $a^{(nk)}$ and $b^{(nk)}$ are known coefficients and $q^{(k)}$ are unknowns to be determined.

For $k = 1, 2, \cdots, N$, the coefficients $a^{(nk)}$, $b^{(nk)}$ and $q^{(k)}$ are given by

$$
a^{(nk)} = \begin{cases}
-\mathcal{D}_1^{(k)}(\overline{x}^{(n)}, \overline{y}^{(n)}, \overline{z}^{(n)}) \\
\qquad \text{if } \phi \text{ is specified over } S^{(k)}, \\
\mathcal{D}_2^{(k)}(\overline{x}^{(n)}, \overline{y}^{(n)}, \overline{z}^{(n)}) - \frac{1}{2}\delta^{(nk)} \\
-\mu_{3D}^{(nk)}\alpha(\overline{x}^{(k)}, \overline{y}^{(k)}, \overline{z}^{(n)}) \\
\qquad \text{if } \partial\phi/\partial n \text{ is specified over } S^{(k)},
\end{cases}
$$

$$
b^{(nk)} = \begin{cases}
\overline{\phi}^{(k)}(-\mathcal{D}_2^{(k)}(\overline{x}^{(n)}, \overline{y}^{(n)}, \overline{z}^{(n)}) + \frac{1}{2}\delta^{(nk)} \\
+\mu_{3D}^{(nk)}\alpha(\overline{x}^{(k)}, \overline{y}^{(k)}, \overline{z}^{(k)})) \\
\qquad \text{if } \phi \text{ is specified over } S^{(k)}, \\
\overline{p}^{(k)}\mathcal{D}_1^{(k)}(\overline{x}^{(n)}, \overline{y}^{(n)}, \overline{z}^{(n)}) \\
\qquad \text{if } \partial\phi/\partial n \text{ is specified over } S^{(k)},
\end{cases}
$$

$$
q^{(k)} = \begin{cases}
\overline{p}^{(k)} & \text{if } \phi \text{ is specified over } S^{(k)}, \\
\overline{\phi}^{(k)} & \text{if } \partial\phi/\partial n \text{ is specified over } S^{(k)},
\end{cases} \tag{6.46}
$$

while, for $j = 1, 2, \cdots, L$, the coefficients $a^{(n[N+j])}$ and $q^{(N+j)}$ are given by

$$
\begin{aligned}
a^{(n[N+j])} &= -\delta^{(n[N+j])} - \mu_{3D}^{(n[N+j])}\alpha(\overline{x}^{(N+j)}, \overline{y}^{(N+j)}, \overline{z}^{(N+j)}) \\
q^{(N+j)} &= \overline{\phi}^{(N+j)}.
\end{aligned} \tag{6.47}
$$

Note that, for $n = 1, 2, \cdots, N + L$ and $m = 1, 2, \cdots, N + L$, we define

$$
\delta^{(nm)} = \begin{cases}
0 & \text{if } n \neq m, \\
1 & \text{if } n = m.
\end{cases}
$$

6.4.2 Implementation on Computer

The subroutine CEDRHZT and its supporting subprograms RHO, CHI, DCHI and CFMU, as listed in Section 3.3, are modified and renamed here as CEDRHZT3D, RHO3D, CHI3D, DCHI3D and CFMU3D respectively, in order to code the three-dimensional dual-reciprocity boundary element method for solving Eqs. (6.2) and (6.37).

The subprograms RHO3D, CHI3D, DCHI3D and CFMU3D which compute respectively the radial basis function ρ_{3D} in Eq. (6.39), χ_{3D} as given in Eq. (6.40), $\partial\chi_{3D}/\partial n$ and $\mathcal{D}_1^{(k)}(\overline{x}^{(n)}, \overline{y}^{(n)}, \overline{z}^{(n)})$, $\mathcal{D}_2^{(k)}(\overline{x}^{(n)}, \overline{y}^{(n)}, \overline{z}^{(n)})$ and $\mu_{3D}^{(nj)}$ defined in Eqs. (6.30) and (6.42) are listed below. Note that the subroutine CPD3D is called in CFMU3D.

```
      function RHO3D(x,y,z,a,b,c)

      double precision RHO3D,x,y,z,a,b,c,r

      r=dsqrt((x-a)**2d0+(y-b)**2d0+(z-c)**2d0)
      RHO3D=1d0+r**2d0+r**3d0

      return
      end


      function CHI3D(x,y,z,a,b,c)

      double precision CHI3D,x,y,z,a,b,c,r

      r=dsqrt((x-a)**2d0+(y-b)**2d0+(z-c)**2d0)
      CHI3D=(r**2d0)/6d0+(r**4d0)/20d0+(r**5d0)/30d0

      return
      end


      function DCHI3D(x,y,z,a,b,c,nx,ny,nz)

      double precision DCHI3D,x,y,z,a,b,c,nx,ny,nz,r

      r=dsqrt((x-a)**2d0+(y-b)**2d0+(z-c)**2d0)
      DCHI3D=(1d0/3d0+(r**2d0)/5d0+(r**3d0)/6d0)
```

```
   & *((x-a)*nx+(y-b)*ny+(z-c)*nz)

      return
      end

      subroutine CFMU3D(N,L,xm,ym,zm,nx,ny,nz,fa,mu)

      integer N,L,k,NL,l1,nn,mm,j

      double precision xm(1000),ym(1000),zm(1000),
   & nx(1000),ny(1000),nz(1000),fa(1000,1000,2),
   & mu(1000,1000),PF1,PF2,comega(1000,1000),
   & cc(1000,1000),RHO3D,omega(1000,1000),
   & PSI(1000,1000),lam,CHI3D,DCHI3D

      NL=N+L

      do 10 k=1,N
      do 10 nn=1,NL
      call CPD3D(xm(nn),ym(nn),zm(nn),k,
   &  nx(k),ny(k),nz(k),PF1,PF2)
      fa(k,nn,1)=PF1
      fa(k,nn,2)=PF2
   10 continue

      do 20 k=1,NL
      comega(k,k)=1d0
      do 20 nn=1,NL
      if (k.ne.nn) comega(k,nn)=0d0
      cc(k,nn)=RHO3D(xm(nn),ym(nn),zm(nn),
   &  xm(k),ym(k),zm(k))
   20 continue

      do 30 j=1,NL
      if (j.eq.1) then
      l1=1
      else
```

```
      11=0
      endif
      call SOLVER(cc,comega(1,j),NL,11,omega(1,j))
 30   continue

      do 40 nn=1,NL
      if (nn.le.(N)) then
      lam=0.5d0
      else
      lam=1d0
      endif
      do 40 mm=1,NL
      PSI(nn,mm)=lam*CHI3D(xm(nn),ym(nn),zm(nn),
     & xm(mm),ym(mm),zm(mm))
      do 40 k=1,N
      PSI(nn,mm)=PSI(nn,mm)
     & -CHI3D(xm(k),ym(k),zm(k),xm(mm),ym(mm),zm(mm))
     & *fa(k,nn,2)
     & +DCHI3D(xm(k),ym(k),zm(k),xm(mm),ym(mm),zm(mm),
     & nx(k),ny(k),nz(k))*fa(k,nn,1)
 40   continue

      do 50 nn=1,NL
      do 50 j=1,NL
      mu(nn,j)=0d0
      do 50 mm=1,NL
      mu(nn,j)=mu(nn,j)+omega(mm,j)*PSI(nn,mm)
 50   continue

      return
      end
```

The main subroutine here is CEDRHZT3D (as listed below) which returns the dual-reciprocity boundary element solution of the problem under consideration. The required inputs are the integer variable N (number of boundary elements which is also the number of collocation points on the boundary), L (number of interior collocation points), the real arrays alpha(1:N+L) and gc(1:N+L) (the values of the coefficients α and g in Eq. (6.37) at all the collocation points; for example, the variable alpha(10) gives the value of α at the 10-th collocation point), the real arrays xm(1:N+L), ym(1:N+L) and zm(1:N+L) (coordinates of the collocation points), the real arrays nx(1:N), ny(1:N) and nz(1:N) (the components of the unit normal vector to the boundary elements), and BCT(1:N) and BCV(1:N) are for recording the boundary conditions as explained on page 39. The output is the real array phi(1:N+L) which gives the numerical value of the solution ϕ at all the collocation points. Note that the subroutine SOLVER is called in CEDRHZT3D.

```
      subroutine CEDRHZT3D(N,L,alpha,gc,xm,ym,zm,
     & nx,ny,nz,BCT,BCV,phi)

      integer N,L,k,NL,nn,mm,j,BCT(1000)

      double precision xm(1000),ym(1000),zm(1000),
     & gc(1000),nx(1000),ny(1000),nz(1000),
     & BCV(1000),phi(1000),A(1000,1000),alpha(1000),
     & B(1000),d1,Z(1000),mu(1000,1000),
     & fa(1000,1000,2)

      call CFMU3D(N,L,xm,ym,zm,nx,ny,nz,fa,mu)

      NL=N+L

      do 60 nn=1,NL
      do 60 k=1,NL
      A(nn,k)=0d0
   60 continue

      do 80 nn=1,NL
      do 70 k=1,N
      if (BCT(k).eq.0) then
      A(nn,k)=-fa(k,nn,1)
```

```fortran
      else
      if (k.eq.nn) then
      d1=0.5d0
      else
      d1=0d0
      endif
      A(nn,k)=fa(k,nn,2)-d1-mu(nn,k)*alpha(k)
      endif
70    continue
      do 75 j=1,L
      if (nn.eq.(N+j)) then
      d1=1d0
      else
      d1=0d0
      endif
      A(nn,N+j)=-d1-mu(nn,N+j)*alpha(N+j)
75    continue
80    continue

      do 180 nn=1,NL
      B(nn)=0d0
      do 170 k=1,N
      if (BCT(k).eq.0) then
      if (k.eq.nn) then
      d1=0.5d0
      else
      d1=0d0
      endif
      B(nn)=B(nn)+BCV(k)*(-fa(k,nn,2)
     & +d1+mu(nn,k)*alpha(k))
      else
      B(nn)=B(nn)+BCV(k)*fa(k,nn,1)
      endif
170   continue
      do 173 j=1,NL
      B(nn)=B(nn)-mu(nn,j)*gc(j)
173   continue
```

```
 180 continue

     call solver(A,B,NL,1,Z)

     do 190 k=1,N
     if (BCT(k).eq.0) then
     phi(k)=BCV(k)
     else
     phi(k)=Z(k)
     endif
 190 continue

     do 200 k=1,L
     phi(N+k)=Z(N+k)
 200 continue

     return
     end
```

Example 6.3

For a test problem, take the solution domain to be the cuboid region $0 < x < 1$, $0 < y < 1$, $0 < z < 1$. We use the dual-reciprocity boundary element method to solve numerically the partial differential equation

$$\frac{\partial^2 \phi}{\partial x^2} + \frac{\partial^2 \phi}{\partial y^2} + \frac{\partial^2 \phi}{\partial z^2} + \frac{4}{4 + 4x^2 + 2y^2 + z^2}\phi$$
$$= \frac{9}{2} \text{ in the cuboid region,}$$

subject to the boundary conditions

$$\phi = \begin{cases} 1 + y^2/2 + z^2/4 & \text{for } x = 0,\ 0 < y < 1,\ 0 < z < 1, \\ 2 + y^2/2 + z^2/4 & \text{for } x = 1,\ 0 < y < 1,\ 0 < z < 1, \\ 1 + x^2 + z^2/4 & \text{for } y = 0,\ 0 < x < 1,\ 0 < z < 1, \\ 3/2 + x^2 + z^2/4 & \text{for } y = 1,\ 0 < x < 1,\ 0 < z < 1, \end{cases}$$

$$\frac{\partial \phi}{\partial n} = \begin{cases} 0 & \text{for } z = 0, \ 0 < x < 1, \ 0 < y < 1, \\ 1/2 & \text{for } z = 1, \ 0 < x < 1, \ 0 < y < 1. \end{cases}$$

The exact solution of the test problem is

$$\phi(x, y, z) = 1 + x^2 + \frac{1}{2}y^2 + \frac{1}{4}z^2.$$

The six faces of the cuboid are discretized into $12N_0^2$ triangular elements as described on page 219. Once the vertices of each the triangular element are set up, the subroutine **BE3D** is called to compute, among other things, the boundary collocation points given by $(\overline{x}^{(j)}, \overline{y}^{(j)}, \overline{z}^{(j)})$ for $j = 1, 2, \cdots, 12N_0^2$. We choose N_1^3 well spaced out collocation points in the interior of the cuboid region. The interior collocation points are given by $([2k-1]/(2N_1), [2m-1]/(2N_1), [2n-1]/(2N_1))$ for $k = 1, 2, \cdots, N_1$, $m = 1, 2, \cdots, N_1$ and $n = 1, 2, \cdots, N_1$. The coordinates of the interior collocation points are stored in the real arrays `xm(N+1:N+L)`, `ym(N+1:N+L)` and `zm(N+1:N+L)`.

The main program **EX6PT3** for the test problem here is listed below.

```
program EX6PT3

integer N0,BCT(1000),N,i,j,ians,ie,N1,k,L

double precision BCV(1000),phi(1000),pint,
& s1,s2,dl,nx(1000),ny(1000),nz(1000),
& alpha(1000),gc(1000),xm(1000),ym(1000),
& zm(1000),xt(1000,3),yt(1000,3),zt(1000,3)

print*,'Enter N0 (for triangular elements):'
read*,N0
print*,'Enter N1 (for interior
& collocation points):'
read*,N1

N=12*N0*N0
L=N1**3d0
```

```
      dl=1d0/dfloat(2*N0)
      ie=-2

      do 10 i=1,N0
      do 10 j=1,N0
      ie=ie+2
      s1=dfloat(2*i-1)*dl
      s2=dfloat(2*j-1)*dl
      xt(ie+1,1)=0d0
      yt(ie+1,1)=s1-dl
      zt(ie+1,1)=s2-dl
                :
                :
                :
```

The complete code for setting up the vertices of
the triangular elements are given in the listing of
EX6PT1 on page 219 .

```
                :
                :
                :
      xt(10*N0*N0+ie+2,3)=s1-dl
      yt(10*N0*N0+ie+2,3)=s2+dl
      zt(10*N0*N0+ie+2,3)=1d0
   10 continue

      call BE3D(N,xt,yt,zt,xm,ym,zm,nx,ny,nz)

      do 30 i=1,N
      if (i.le.(8*N0*N0)) then
      BCT(i)=0
      BCV(i)=1d0+xm(i)**2d0+0.5d0*ym(i)**2d0
     & +0.25d0*zm(i)**2d0
      else
      BCT(i)=1
      BCV(i)=0.5d0*zm(i)
```

```
      endif
30 continue

   dl=1d0/dfloat(N1)
   ie=0

   do 35 i=1,N1
   do 35 j=1,N1
   do 35 k=1,N1
   ie=ie+1
   xm(12*N0*N0+ie)=dfloat(2*i-1)*dl*0.5d0
   ym(12*N0*N0+ie)=dfloat(2*j-1)*dl*0.5d0
   zm(12*N0*N0+ie)=dfloat(2*k-1)*dl*0.5d0
35 continue

   do 40 i=1,N+L
   alpha(i)=4d0/(4d0+4d0*xm(i)**2d0+2d0*ym(i)**2d0
 & +zm(i)**2d0)
   gc(i)=4.5d0
40 continue

   call CEDRHZT3D(N,L,alpha,gc,xm,ym,zm,
 &    nx,ny,nz,BCT,BCV,phi)

   print*,' x y z Numerical Exact'
   print*,'--------------------'

   do 50 i=N+1,N+L
   write(*,60) xm(i),ym(i),zm(i),phi(i),
 & 1d0+xm(i)**2d0+0.5d0*ym(i)**2d0
 & +0.25d0*zm(i)**2d0
50 continue

60 format(5F10.6)

   end
```

The main program EX6PT3 is compiled together with the subprograms BE3D, CEDRHZT3D, CFMU3D, CHI3D, CPD3D, DCHI3D, DFS3D, FS3D, RHO3D and SOLVER (together with its supporting subprograms) into an executable code. The code is used to obtain numerical values of ϕ at selected collocation points in the interior of the cuboid region. In Table 6.2, the numerical values of ϕ obtained using $N_0 = 5$ (300 boundary elements) and $N_1 = 2$ (8 interior collocation points) are compared with the exact solution. Even for a relatively crude discretization of the boundary and a low number of interior collocation points, the computation is reasonably accurate, as the percentage error of each of the numerical values is below 1%.

Table 6.2

(x, y, z)	DRBEM	Exact
$(0.25, 0.25, 0.25)$	1.108176	1.109375
$(0.25, 0.25, 0.75)$	1.232155	1.234375
$(0.25, 0.75, 0.75)$	1.491175	1.484375
$(0.75, 0.25, 0.25)$	1.610467	1.609375
$(0.75, 0.25, 0.75)$	1.732215	1.734375
$(0.75, 0.75, 0.25)$	1.870795	1.859375
$(0.75, 0.75, 0.75)$	1.994211	1.984375

6.5   Summary and Discussion

We have shown how the boundary element procedures in Chapters 1 and 3 can be extended to include three-dimensional problems. The extension is theoretically straightforward, but its practical implementation is a more involved task.

The boundary of the solution domain has to be discretized into surface elements. In general, it appears most convenient to discretize the surface boundary into triangular planar elements. Thus, boundary element procedures involving triangular elements are given in this chapter.

We approximate the required solution and its normal derivative on the boundary as constants over each of the elements. For three-dimensional problems, it may also be possible to develop discontinuous linear elements similar to the ones in Chapter 2.

The fundamental solution involved and its normal derivative on the boundary have to be integrated over every triangular element. The surface integrals over a triangular element are first transformed into double integrals over a square domain. We then use an accurate Gaussian integration formula to compute numerically the double integrals. Under certain situations, such as when the coefficient w in Eq. (6.34) has a relatively large magnitude, it may be necessary to refine the numerical integration over the square domain.

6.6   Exercises

1. Modify the program **EX6PT1** in Example 6.1 for solving numerically the three-dimensional Laplace's equation in the cuboid region $0 < x < 1,\ 0 < y < 1,\ 0 < z < 1$, subject to the boundary conditions

$$\phi = \begin{cases} 1 & \text{on the face } z = 1,\ 0 < x < 1,\ 0 < y < 1, \\ 0 & \text{on the remaining five faces of the cuboid.} \end{cases}$$

Compute ϕ at selected points in the interior of the cuboid region and compare the numerical values obtained with the series solution[4]

$$\phi(x, y, z)$$
$$= \frac{16}{\pi^2} \sum_{m=1}^{\infty} \sum_{n=1}^{\infty} \frac{\sin([2m-1]\pi x)\sin([2n-1]\pi y))}{(2m-1)(2n-1)}$$
$$\times \frac{\sinh(\pi\sqrt{(2m-1)^2 + (2n-1)^2}\,z)}{\sinh(\pi\sqrt{(2m-1)^2 + (2n-1)^2})}.$$

---

[4]This series solution may be derived by using separation of variables and double Fourier series.

2. Modify the program **EX6PT3** in Example 6.3 for the numerical solution of the boundary value problem in Example 6.2. Compare the numerical values of ϕ at the interior collocation points with the exact solution $\phi(x, y, z) = \sqrt{2}\sin(\pi z/4)$.

3. Consider the task of solving the three-dimensional Laplace's equation outside a given closed surface S subject to the conditions

$$
\begin{aligned}
\phi &= f(x, y, z) \text{ for } (x, y, z) \in S, \\
\phi &\to 0 \text{ as } x^2 + y^2 + z^2 \to \infty,
\end{aligned}
$$

where f is a suitably prescribed function of x, y and z. How would you modify the computer codes in Subsection 6.2.7 for the numerical solution of the exterior boundary value problem? Test your computer codes on the special case in which S is the spherical surface $x^2 + y^2 + z^2 = 1$ and $f(x, y, z) = z$. (Note. The exact solution for the special case is $\phi = z/(x^2 + y^2 + z^2)^{3/2}$.)

4. For the three-dimensional Laplace's equation in the half space $z > 0$, use the method of image to construct a Green's function $\Phi(x, y, z; \xi, \eta, \zeta)$ which satisfies the boundary condition

$$
\Phi(x, y, 0; \xi, \eta, \zeta) = 0.
$$

(Note. Let $\Phi(x, y, z; \xi, \eta, \zeta)$ take the form

$$
\Phi(x, y, z; \xi, \eta, \zeta) = \Phi_{3D}(x, y, z; \xi, \eta, \zeta) + \Phi^*(x, y, z; \xi, \eta, \zeta),
$$

where $\Phi_{3D}(x, y, z; \xi, \eta, \zeta)$ is the fundamental solution defined in Eq.(6.6). Choose $\Phi^*(x, y, z; \xi, \eta, \zeta)$ in such a way that the required boundary condition is satisfied. For a guide, refer to Section 5.2 on page 157.)

5. Repeat Exercise 4 for the boundary condition

$$
\frac{\partial}{\partial z}[\Phi(x, y, z; \xi, \eta, \zeta)]\bigg|_{z=0} = 0.
$$

6. If the Green's function in Exercise 4 is used in place of the fundamental solution $\Phi_{3D}(x, y, z; \xi, \eta, \zeta)$ in Eq.(6.6), how would you modify the main program **EX6PT1** and the subroutines **CE3D1** and **CE3D2** to solve the boundary value problem in Exercise 1 with fewer boundary elements?

Index